STATISTICAL REASONING
AND DECISION MAKING

PIERRE SIMON DE LAPLACE
1749-1827

Author of *Theorie Analytique des Probabilités*

STATISTICAL REASONING AND DECISION MAKING

Enders A. Robinson
Visiting Professor of Geophysics
Cornell University

Goose Pond Press
11600 Southwest Freeway, Suite 179
Houston, Texas 77031

Statistical Reasoning
and Decision Making

First Printing 1981

Library of Congress Catalog Card Number 81-85240
    Robinson, Enders Anthony
        Statistical Reasoning and Decision Making
    Houston: Goose Pond Press
        200 p.
8112   811027

Dedicated to

HERMAN WOLD
Professor of Statistics
Uppsala University
Uppsala, Sweden

CONTENTS

Preface   xi

1. Frequency distributions   1
   History   1
   Collection of numerical data   1
   Exercises   5

2. Mean, variance, and standard deviation   7
   Mean   7
   Variance   8
   Standard deviation   11
   Standard scores   11
   Exercises   12

3. Probability   13
   Random processes   13
   Statistical independence   14
   Classical definition of probability   15
   Relative frequency definition of probability   16
   Personal, or subjective, definition of probability   18
   Events   18
   Expectation of an event   20
   Expectation of several events   21
   Advantageous, equitable, and disadvantageous games   22
   Betting on events   23
   Application to life insurance   23
   Exercises   23

4. Rules for probabilities   25
   Addition rule for probabilities   25
   Multiplication rule for probabilities   29
   Tree diagrams   33
   Exercises   36

5. Binomial distribution   37
   Family make-up   37
   Pascal's triangle   39
   Permutations and combinations   40
   The binomial distribution   42
   Mean of the binomial distribution   43
   Variance of the binomial distribution   45
   Independence between individual births   46
   Exercises   48

6. Population and sample   51
   Population   51
   Sample   52
   Random sampling   52
   Theoretical frequency curve (for a finite population)   53
   Theoretical frequency curve (for an infinite population)   54
   Exercises   56

7. The normal distribution   59
   Background   59
   The central limit theorem and the normal curve   59
   The standard normal distribution   61
   Exercises   63

8. Expectation   69
   Introduction   69
   Properties of expectation   69
   Variance   70
   Random sample   72
   Sample variance   74
   Unbiased estimates   75
   Exercises   76

9. Entropy   79
   Definition of entropy   79
   Simple examples   80
   Entropy as a measure of uncertainty   83
   Exercises   84

10. The uncertainty principle   87
   Basic idea of the uncertainty principle   87
   Application of the uncertainty principle   88
   Interference or interaction patterns   90
   Exercises   92

11. Decision making   95
   Null hypothesis, significance, and the level of significance   95
   Type I and Type II Errors   97
   Exercises   99

12. Bayesian decision making   101
   Bayes rule for flipping trees   101
   Reaction to experimental evidence   104
   Hypothesis testing   106
   Exercises   113

13. Analysis of decision under uncertainty   115
    Conditional analysis and the payoff table   115
    Possible criteria for selecting an act   119
    The expected monetary value criterion   119
    Assessment of probabilities   121
    Probabilities based on relative frequencies   123
    Judgmental weights   125
    The value of information   127
    Expected utility and attitude toward risk   130

14. Background in language   137
    The language of Wall Street   137
    Probability concepts in everyday language   139
    Use of language   145

15. Stock market speculation as a game   147
    Model building   147
    Model for the stock market   149
    Plight of the small investor   154

Appendix   157
    Review exercises   157
    One hundred review problems   161
    Greek alphabet   175
    Table I: Binomial coefficients   176
    Binomial probabilities for $p = 0.5$ 177
    Table II: Binomial probabilities   178
    Table III: Normal curve areas between 0 and z 180
    Table IV: Normal curve areas beyond z 181
    Table V: Random numbers   182

Bibliography   183

Books by Enders A. Robinson   185

Index   186

PREFACE

People in business, industry, the professions, and government constantly face the task of making decisions based upon incomplete information. The need for decision making is always present. Each time a person makes a decision he faces the risk that the decision may be wrong. People who have the ability to identify relevant facts and the skill to analyze the problem have a better chance to reach the right decision.

Statistics (plural) are numerical data that serve as a record of past actions. Statistics (singular) is the science of the collection, analysis, and interpretation of numerical data. In recent years, tremendous strides have been made in developing more powerful statistical techniques for the analysis of decision problems. These methods are referred to collectively as statistical decision theory. Sound and accurate statistical reasoning is required in the process of decision making. The foundation of statistics is mathematics, particularly probability theory; its methodology is scientific; and its focus is on problem solving.

This book is intended for people who want a mathematically sound, but elementary introduction to statistical reasoning and decision making. The theory of probability, as the foundation upon which the methods of statistics are based, is treated early in the book. Many textbooks on probability and statistics are written for readers who have the mathematical sophistication that comes from a working knowledge of calculus. However, it is just as worthwhile to bring the essential elements of this important subject to the attention of those readers who do not have a calculus background. Such is the path which has been taken in this book. Our purpose is to give the reader a sense of the nature and achievements of probability and statistics without making use of calculus.

This book is not intended to be the main text of a college course in statistics. Instead it can be used as a supplementary text, and preferably it should be assigned before work on the main text begins. Used in this way, it can be immensely helpful in breaking the psychological barrier that often stands between students and statistics, especially if statistics is a required course for those students. The most frequent use of the book, however, will probably be as basic reading in science courses that require some statistical knowledge, such as in economics and in geology. The chapters of the book do not have to be read in order, and many readers will select those chapters which are of the most interest to them.

It is a book for beginners in statistics. Emphasis is on the need to understand basic concepts, and not so much on the manipulation of the data. In this context, many of the standard topics are not covered at all, such as t-tests and analysis of variance. Instead more conceptual topics such as mathematical expectation, entropy, and the uncertainty principle are discussed. Once the reader understands basic concepts, then he is better prepared to become a practitioner of statistics, if he wishes, where data processing methods are essential. He can then turn profitably to the many excellent books on statistical methodology and computation.

This book can therefore function effectively in two ways. First it can present some of the basic ideas of statistics in a form that is helpful to people in other disciplines.

Secondly, for those who want a more advanced training in statistics, it can provide a useful framework for the more detailed study that lies ahead. Many problems and exercises are given so that the reader can test and develop his understanding of the subject matter as he progresses. The presentation in this book emphasizes not the facts of statistics as such, but the type of phenomena and circumstances, the method of proposing problems, and the method of solving problems.

Enders A. Robinson

Ithaca, New York
December 21, 1981

CHAPTER 1

FREQUENCY DISTRIBUTIONS

History

One of the first recorded uses of statistical reasoning was made by the Chinese over 2200 years ago. By means of statistical data collected by censuses of their population, it was observed that the ratio of male births to female births remained almost unchanged from year to year, and that this ratio was approximately one to one.

In 1662 John Graunt wrote a book entitled *Natural and Political Observations upon Bills of Mortality*. The Bills of Mortality were lists of births and deaths in London which were made at weekly intervals and the records went back to the great plague of 1603. Graunt found that there was a rough balance between the sexes, with male births slightly exceeding female. This was a novel idea at a time when it was commonly believed that there were three women to every man. In addition Graunt constructed the first "expectation of life" table which links the death rate to the number of survivors at various ages. In 1690, Sir Edmund Halley, the Astronomer Royal who identified what we call Halley's Comet by means of a computed average, constructed an improved life table. The life table provided the foundation of the life insurance industry, and the first life insurance company to operate on a statistical basis, the Equitable, was founded in London in 1762.

It was 1791 when the word "statistics" first appeared in the English language. It was used by Sir John Sinclair in the title of his book *A Statistical Account of Scotland* which was based on materials gathered from the parish ministers in every county. He borrowed the word from German, where statistics was a word referring to the political science dealing with state affairs. Sinclair altered its meaning to the sense that we use the word today, namely a collection of numerical data. In his preface he states: "As I thought a new word might attract more public attention, I resolved to adopt it."

Collection of numerical data

This chapter is concerned with the problems of making a summary of numerical data. The kind of summary that we will use is the frequency distribution. The following table gives the United States Population Distribution by age for 1960:

1

FREQUENCY DISTRIBUTION OF AGE
(1960 U.S. Census)

Under 5	5-19	20-44	45-64	65 and over	Total
20,321	48,798	58,251	36,076	16,559	180,005

(Frequencies given in thousands of people)

Let us now look at the characteristics of this table. The age of a person represents a numerical variable. A numerical variable is one whose values have a necessary order and numerical relation to each other. For example, the variable sex (which takes the values male or female) is not a numerical variable, so people classified according to sex fall into unordered classes. On the other hand people classified according to age fall into ordered classes. The records collected by the U.S. Census were for age at the last birthday, and in this table for convenience of presentation the original data have been grouped into the intervals: Under 5, 5-19, 20-44, 45-64, 65 and over. The first interval is one of 5 years, the second 15 years, the third 25 years, the fourth 20 years, and the highest interval is an open interval (i.e. one for which no upper limit is stated). Every person in the Census is allocated to one of these intervals which are called class intervals. The 180,007,000 persons enumerated in the 1960 census have been classified into these age groups. These classes are mutually exclusive (because no person can be in more than one of them) and exhaustive (because no person can be in none of them).

The above table is an example of a frequency distribution. A *frequency distribution* is made up of (1) a set of mutually exclusive and exhaustive classes and (2) the number of individuals belonging to each class. The number of individuals in any given class is called a *frequency*. Because the frequencies in all the classes add up to the total, we can divide each frequency by the total to obtain the *relative frequency distribution*. For example, the relative frequency distribution derived from the above table is:

RELATIVE FREQUENCY DISTRIBUTION OF AGE
(1960 U.S. Census)

Under 5	5-19	20-44	45-64	65 and over	Total
11.3	27.1	32.4	20.0	9.2	100.0

(Relative frequencies given in percent)

There are many different ways of picturing a frequency distribution. We will discuss here the histogram, the frequency polynomial, and the frequency curve. At this point we would like to make use of the conventional mathematical terms *abscissa* and *ordinate*. Let a point P be located in reference to two axes at right angles to each other, as shown in the Figure. Then the horizontal distance from the vertical axis to the point is called the abscissa, and the vertical distance from the horizontal axis to the point is called the ordinate. Together the abscissa and the ordinate are called the coordinates of the point.

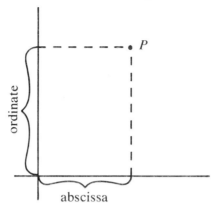

Let us now prepare a histogram for the relative frequency distribution of age. Let the horizontal axis represent the age scale. Mark the abscissas corresponding to the division points between intervals. We obtain

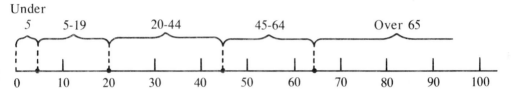

The first thing we notice is that each class interval is of a different length. This fact makes our problem of constructing a histogram more difficult, but it will illustrate the important point that the frequencies correspond to areas of the histogram, and not to ordinates. Thus over the class interval "Under 5" we wish to place a rectangular block whose area is proportional to 11.3. Let that block be given by a width of the class interval (5 years), times a height of 11.3 cm. Over the class interval 5-19 we wish to place a rectangular block whose area is proportional to 27.1. Since the width of this class interval (15 years) is three times the width of the first class interval, we must divide 27.1 by 3 to obtain the height of the required block, i.e. $27.1 \div 3 = 9.03$ cm. Over the class interval 20-44 we wish to place a rectangular block whose area is proportional to 32.4. Since the width of this class interval (25 years) is five times the width of the first class interval, we obtain the height of the required block to be $32.4 \div 5 = 6.48$ cm. The height of the required block for the next to last class interval is $20.0 \div 4 = 5.00$ cm. The last class interval presents a problem because it is an open interval. However, we could certainly close the interval at 100 years, because of the very few people above that age. Because of the relatively few number of people over 90 years old, let us instead close it at 90 years, so it has a length of 25 years. Under this assumption the height of the required block for the last class interval is $9.2 \div 5 = 1.84$ cm.

(Decisions like closing the last interval at 90 years often have to be made in classifying data; it is a good policy to explain in footnotes how such difficulties are resolved on any statistical table or plot.) The final result is the histogram shown in Figure 1.1.

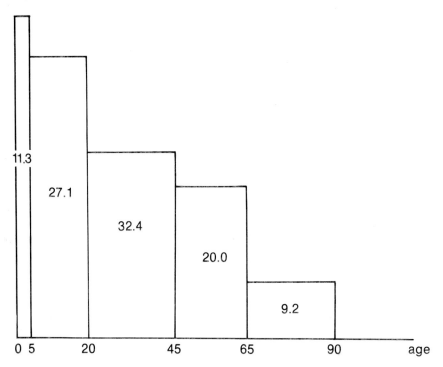

Figure 1.1. Histogram of the percent distribution of age in the 1960 U.S. Census (Note: Ages 65 and over lumped into interval 65-90).

For a frequency polygon the horizontal and vertical scales are laid off exactly as for a histogram. For each interval a point is located at the top of each block directly above the middle of the class interval. In other words the abscissa of the point is the age at the middle of the interval and its ordinate is the height of the block. Note that if an interval has zero frequency the point representing that block will lie on the horizontal axis. For our example, let us arbitrarily assign such a zero frequency point at age 100 years. The frequency polygon for the histogram of Figure 1.1 is shown in Figure 1.2.

Finally the frequency curve is the smooth curve drawn through the frequency polygon as shown in Figure 2. The histogram is the most detailed picture of the frequency distribution; the frequency polygon tries to smooth out the rough block-like appearances of the histogram; the frequency curve is the artistic brush-stroke that hopefully reduces the data to its basic features. In our case we used the frequency curve to smooth out a depression in the frequency polygon

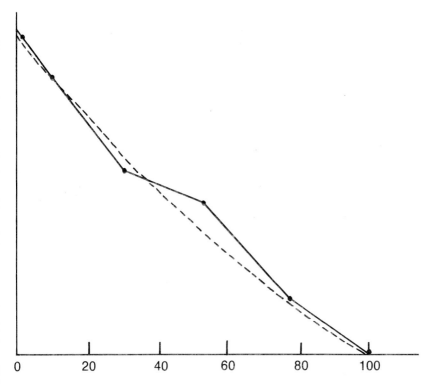

0   20   40   60   80   100

Figure 1.2. Frequency polygon (solid curve) and frequency curve (dashed curve) for the histogram shown in Figure 1.1.

that appeared in the class interval 20-44 years. Whether such a smoothing is justified or not can only be determined by examination of more detailed data.

Exercises

1. The U.S. Census for 1960 gave the following frequency distribution. Age Groups: Under 5, 5-9, 10-14, . . . , 80-84, 85 years and over. Corresponding Frequencies (in thousands): 20321, 18691, 16773, 13334, 11063, 10981, 12026, 12541, 11640, 10893, 9610, 8431, 7142, 6258, 4739, 3053, 1580, 929 respectively. Find the relative frequency distribution and plot the histogram, frequency polynomial, and frequency curve. Is there a depression on the frequency polynomial in the neighborhood of 25 years? Explain in terms of the 1930-1940 depression.

2. In order to avoid ambiguity in constructing a frequency distribution, we select class boundaries so that there can be no question as to which class a given observation belongs. One way to assign boundaries is the way we assign ages: A person 19 years 364 days old is called 19 years. Using this method suppose the class intervals are 1-7, 8-14, 15-21, 22-28, which we call intervals I, II, III, IV

respectively. To which interval would the score 7.89 be assigned? The score 14.99? The score 28.001? The score 7? The score 8? The score 15?

3. Using the method in question 2, suppose the class interval for a B was 80-89, and for an A, 90-100. Suppose you got an 89.9. Would you get an A or B for the course?

4. A histogram by construction has a total area equal to the total number of data points or one hundred percent. In the case of equal class intervals, show that the construction of the frequency polygon as described in the text guarantees that both histogram and frequency polygon have the same area, provided that a class with zero frequency is added at each end.

5. Does a frequency polygon in the case of unequal class intervals have the same area as the corresponding histogram?

6. Construct two frequency distributions from the scores of 50 students given by 56, 48, 48, 48, 44, 54, 46, 54, 50, 41, 46, 68, 46, 54, 51, 44, 40, 31, 46, 42, 55, 46, 58, 40, 43, 45, 43, 50, 26, 48, 54, 31, 38, 44, 52, 36, 35, 56, 56, 50, 43, 37, 52, 56, 26, 60, 48, 50, 54, 56. Let the first distribution be one with equal intervals of 3 units each starting with 0-2, 3-5, 6-8, . . . Let the second distribution be one with equal intervals of five units each starting with 0-4, 5-9, 10-14, . . . Draw the histogram and frequency curve for each distribution, and comparing the two curves decide which of the two distributions gives a clearer presentation of the data.

7. From each of the two distributions obtained in Exercise 6, find (a) the number of students with scores between 30 and 45, (b) the number of students whose scores were at least 45. Explain why the two distributions give different answers, and find the exact answers from the original data.

8. Draw the frequency polygon and frequency curve for the distribution of the length of the left occipital bone in millimeters of old Egyptian skulls given by: Class intervals: 84-85, 86-87, 88-89, . . . , 118-119. Frequencies: 12, 12, 32, 48, 79, 116, 104, 126, 123, 74, 68, 36, 18, 7, 4, 4, 0, 1. (From T. L. Woo, 1930, *Biometrika*, Vol. 22, page 324.)

CHAPTER 2

MEAN, VARIANCE, AND STANDARD DEVIATION

Mean

The frequency distributions encountered in practice can vary considerably in their general shape. Some are symmetrical and some are skew. Some have one peak; others have two or more peaks. Ones with only one peak are called *unimodal*; ones with two peaks are called *bimodal*.

Often we wish to compare two distributions. If one is a unimodal symmetric distribution, and the other is extremely skew, then a concise comparison would be difficult to make, and so we would have to specify both distributions in detail. However often we want to compare two distributions of the same general type, and then it is possible to make a satisfactory comparison by examining only a few principal characteristics. In the case of two unimodal symmetrical distributions it is usually adequate to compare their means and variances. The *mean* is a measure of central value which essentially locates the distribution, whereas the *variance* is a measure of spread which essentially gives the degree of scatter about the mean. Often the standard deviation is used instead of the variance. The *standard deviation* is simply the (positive) square root of the variance.

There are other measures of central value such as the *median* and the *mode*, but the mean is the one most commonly used. Actually, there are three well-known kinds of means, namely the *arithmetic mean*, the *geometric mean*, and the *harmonic mean*. We are concerned with the arithmetic mean which is also called the *arithmetic average*, or simply the *average*. We want to give two formulas for computing the mean: one in the case of raw data and the other in the case of classified data (in the form of a frequency distribution). In the case of raw data the arithmetic mean is found by adding up the data and then dividing by the number of data points. For example, if the data were the scores 60, 80, 70, 85, 65 the mean is the average

$$\frac{60 + 80 + 70 + 85 + 65}{5} = \frac{360}{5} = 72.$$

Let us introduce a few symbols. Let the letter x represent "any score" (i.e. any numerical observation or data point). "The sum of" is customarily denoted by the capital Greek letter sigma Σ. Then Σx means "the sum of all scores." The letter n indicates the number of scores under consideration. The average will be denoted by $\bar{x}$, which is read as x-bar. Using these symbols the formula for the mean can be expressed definitely and concisely by

$$\bar{x} = \frac{\Sigma x}{n}$$

If we had another set of n scores y then the average of these scores would be

$$\bar{y} = \frac{\Sigma y}{n}$$

In case of data in the form of a frequency distribution the arithmetic mean is computed in the following way. Let x denote the center value of any interval, and let f denote the frequency of that interval. Then the arithmetic mean is given by

$$\bar{x} = \frac{\Sigma fx}{n}$$

where n is the total number in all the intervals, that is,

$$n = \Sigma f$$

If the frequency distribution is in terms of relative frequencies

$$f' = f/n$$

then the formula for the arithmetic mean becomes

$$\bar{x} = \Sigma(f/n)x = \Sigma f'x$$

For example, the relative frequency distribution of age for the 1960 U.S. Census given in the last chapter is:

RELATIVE FREQUENCY DISTRIBUTION OF AGE
(1960 U.S. Census)

Interval	Under 5	5-19	20-44	45-64	65 and over	Total
Center Value	2.5	12.5	32.5	55	Say 77.5	
	0.113	0.271	0.324	0.200	0.092	1.000

The arithmetic mean is

$$\bar{x} = \Sigma f'x$$
$$= 0.113(2.5) + 0.271(12.5) + 0.324(32.5) + 0.200(55) + 0.092(77.5)$$

which gives

$$\bar{x} = 32.3$$

We say that the average age of an American in 1960 was 32.3 years.

Variance

When central value is measured by the mean, then variability should be measured by a quantity based on deviations from the mean. Such a quantity is the *variance.* Let us first show how to compute the variance in the case of raw data, such as the five scores given in the preceding section. For each of the five scores 60, 80, 70, 85, 65 (with arithmetic mean $\bar{x} = 72$ as computed previously), we compute the difference of each score from the mean score, that is, we compute $x - \bar{x}$ and in addition perform the computation shown in the following table:

Score	Deviation	Squared Deviation
x	$x - \bar{x}$	$(x - \bar{x})^2$
60	$60 - 72 = -12$	144
80	$80 - 72 = 8$	64
70	$70 - 72 = -2$	4
85	$85 - 72 = 13$	169
65	$65 - 72 = -7$	49
Total 360	0	430

(We note that the sum of the deviations from the means is always identically zero no matter how great or how small the variability; that is,

$$\Sigma(x - \bar{x}) = 0$$

because positive and negative deviations from the mean exactly cancel each other out. Thus this sum provides no useful indication of variability.)

As a measure of variability we would like to have a quantity that is zero only when there is no variability (that is, when all scores are the same) and which becomes larger as the spread among scores is increased. In order that positive and negative deviations do not cancel, we first square each deviation. All of the squared deviations are positive. We then take the sum of squared deviations; for our data the sum of squared deviations is

$$\Sigma(x - \bar{x})^2 = 430$$

However this number reflects not only the variability of the scores but also the number of scores n. As a result we divide by $n - 1$; the result is the variance, denoted by s^2:

$$s^2 = \frac{\Sigma(x - \bar{x})^2}{n - 1}$$

(In Chapter 8 we will give reasons why the denominator of s^2 is $n - 1$ instead of n). For our data the variance is

$$s^2 = \frac{430}{4} = 107.5$$

For the five scores making up our data, the computation of the variance was simple because the mean $\bar{x}$ was an integer and so the deviations $x - \bar{x}$ were also integers. In general, data will not be so well behaved, and the method given will involve large rounding errors unless the deviations are carried to many decimal places. An equivalent formula for the variance which in most cases requires less arithmetic work is

$$s^2 = \frac{n\Sigma x^2 - (\Sigma x)^2}{n(n-1)}$$

For our data we make the table

x	x^2
60	3600
80	6400
70	4900
85	7225
65	4225
Total 360	26350

which gives

$$s^2 = \frac{5(26350) - (360)^2}{5(4)} = \frac{131750 - 129600}{20} = \frac{2150}{20} = 107.5$$

A third formula for the variance is

$$s^2 = \frac{\Sigma x^2 - n\bar{x}^2}{n-1}$$

which for our data gives

$$s^2 = \frac{26350 - 5(72)^2}{4} = \frac{430}{4} = 107.5$$

Let us now show how to compute the variance in the case of classified data (i.e. data in the form of a frequency distribution). The formula for the variance is

$$s^2 = \frac{1}{n-1} \Sigma f(x - \bar{x})^2$$

where n is the total number in all the classes and $\bar{x}$ is the mean:

$$n = \Sigma f, \quad \bar{x} = \frac{\Sigma fx}{n}$$

If the frequency distribution is in terms of relative frequencies $f' = f/n$, then the formula becomes

$$s^2 = \frac{1}{n-1} \Sigma nf'(x - \bar{x})^2 = \frac{n}{n-1} \Sigma f'(x - \bar{x})^2$$

In such cases n is usually a large number so the ratio $n/(n-1)$ is almost equal to one; hence we may then use the formula

$$s^2 = \Sigma f'(x - \bar{x})^2$$

For the relative frequency distribution of age for the 1960 U.S. Census given in the last chapter the values of $x - \bar{x}$ are -29.8, -19.8, 0.2, 22.7, 45.2 so the variance is

$$s^2 = 0.113(-29.8)^2 + 0.271(-19.8)^2 + 0.324(0.2)^2 + 0.200(22.7)^2 + 0.092(45.2)^2$$

which is

$$s^2 = 497.622.$$

Standard deviation

The variance is a measure of the spread of a distribution but no graphic representation can be made of it. Its square root, however, does represent a distance that can be measured along the scale of the scores. As a matter of fact, the standard deviation acts as the standard unit in which to measure deviations of individual scores from the mean. The symbol for the *standard deviation* is s:

$$s = \sqrt{\text{variance}} = \sqrt{s^2} = \sqrt{\frac{\Sigma(x - \bar{x})^2}{n - 1}}$$

For the five scores 60, 80, 70, 85, 65 the standard deviation is

$$s = \sqrt{107.5} = 10.3$$

For the 1960 U.S. Census on age distribution the standard deviation is

$$s = \sqrt{497} = 22.3$$

Standard scores

Suppose that a student received a score of 83 on a mathematics examination and a score of 80 on a language examination. Can these results be interpreted that his standing is about the same on both tests? Without further information we can draw no conclusions because the raw scores do not properly reflect his relative positions on the two tests.

Suppose now that we know that the mathematics average was 75 and the language average was 65. The student stood 8 points above average in mathematics and 15 points above average in language. But can these figures be considered comparable when the scores on the two examinations were not equally variable. If the mathematics standard deviation was 10 and the language standard deviation was 5 then the student was 0.8 standard units above average in mathematics but 3 standard units above average in language. In other words, to allow for differences in the standard deviations in different examinations, the deviation from mean in each examination is divided by the standard deviation of that examination, producing the value

$$z = \frac{x - \bar{x}}{s}$$

This z-value is called the *standard score*. In mathematics the student's standard score was

$$z = \frac{83 - 75}{10} = 0.8$$

whereas in language his standard score was

$$z = \frac{80 - 65}{5} = 3$$

Thus comparatively he did much better in language than mathematics.

Exercises

1. Two variables x and y assume the values

$$x: \quad 6 \quad -8 \quad 10 \quad -3$$
$$y: \quad -8 \quad -5 \quad 4 \quad 2$$

Compute Σx, Σy, Σx^2, Σy^2, Σxy, $(\Sigma x)(\Sigma y)$, $\Sigma(x + y)$, $\Sigma x + \Sigma y$

2. If the variable x is indeed a constant, that is, if x assumes only the constant value 5:

$$x: \quad 5 \quad 5 \quad 5 \quad 5 \quad 5 \quad 5$$

find Σx. If the variable x is a constant with data points

$$x: \quad a \quad a \quad a \quad a \quad a \quad a \quad a$$

then what is n and what is Σx?

3. The grades of 10 students on an exam are:

$$x: \quad 55, \ 85, \ 72, \ 81, \ 83, \ 79, \ 90, \ 68, \ 85, \ 82$$

Find the mean and standard deviation. Then find the standard score for each student.

4. Prove that the sum of deviations $x - \bar{x}$ is zero. Hint: Write $\Sigma(x - \bar{x}) = \Sigma x - \Sigma \bar{x}$, and then use Exercise 2 (as $\bar{x}$ is a constant) to find $\Sigma \bar{x}$.

5. If $x = y + z$ show that the average of x is equal to the sum of the averages of y and z; that is, show $\bar{x} = \bar{y} + \bar{z}$.

6. The IQ's of 500 grade school children are:

Class mark x: 74 78 82 86 90 94 98 102 106 110 114 118 122 126 130
Frequency f: 3 10 15 29 44 72 95 77 55 37 28 19 10   4   2

Find the mean, standard deviation, and the standard scores. Plot the frequency polygon and the frequency curve.

7. Let the x data be 4, 6, 8, 8, 4. Find Σx, Σx^2, $\bar{x}$, s^2. Find the z-score for each data point.

8. Which did a student do better in: A grade of 72 in a course with mean 70 and standard deviation 5 or a grade of 82 in a course with mean 78 and standard deviation 10?

CHAPTER 3

PROBABILITY

Random processes

One identifying feature of a random process is that it has many possible results, or cases. Since we will be speaking about the cases of a random process many times, let us give a name to the collection of all the cases. Since these cases are elemental, exclusive, and exhaustive, it is appropriate to call the set of all cases the *universal set*, or briefly the *universe*, of the random process. Another name for this universal set is the *sample space*.

Using this concept of universe, examples of random process may be described as follows:

Random Process	Universe (or Sample Space)
A coin is tossed.	Heads, Tails.
A die is tossed.	1, 2, 3, 4, 5, 6.
A card is drawn from a pack of 52 cards.	Ace of spades, Ace of clubs, etc., until all the 52 cards are listed.
A rain drop falls on a piece of paper.	The points on the paper.
A seed falls from a tree and lands in a meadow.	The points in the meadow.
An arrow is shot at a target.	The points on the target.

Each outcome of a random process is a happening. It is not an occurrence that can be exactly fixed in advance, perfectly predicted, etc. but instead it is an occurrence that depends in some way and to some extent upon chance, luck, hap, fortune, etc. All that we mean is that there is something about a random process that is uncertain, unsure. This uncertainty as a rule is not entirely unlimited, but only prevails in certain directions and up to certain points.

For example, the statement:

A man attains some age before death.

represents a random process. The age to which a man lives cannot be exactly fixed in advance. No person can calculate what may be the length of any particular life. The hour of a man's death is not predetermined within man's knowledge. When death comes, it happens. This happening is the final outcome of the process of life.

The universe that represents his age consists of all the numbers from 0 as the lower limit to say 150 years as the upper limit, as we can feel perfectly certain that his life will not stretch out more than 150 years. Thus the uncertainty about a man's final age is not entirely unlimited, but can be said to lie within a universe extending from 0 to 150 years.

We toss a penny into the air. It happens to fall heads or it happens to fall tails. This falling is the outcome of this process of gambling. Its universe is *heads, tails,* so that the uncertainty here is limited to two cases, or three if we include the possibility of its landing on edge.

Suppose that we measure the height to which a (normal) man grows. His height will of course lie between certain extremes which delineate the universe, say between 1 meter and 3 meters. The height to which he grows represents a process of life. His height is a happening, that is, an occurrence dependent to some extent upon unknown and unobtainable factors.

As another example, consider the sex of a newborn baby. The baby may happen to be a boy or happen to be a girl. Its sex represents the outcome of the process of birth. The universe consists of two cases, namely *male, female.*

One point should be made clear. Some of the previous examples should have brought this point out, but anyhow it is worth stressing. The point is this: The role played by chance, luck, hap, fortune, etc., in a random process is not required to be complete but may be to any degree. For example, when a coin is tossed or a die is cast, we usually regard the role played by chance as being entire, and the role played by the person as having no effect on the ultimate result. However there may be individuals with extraordinary skill who can control some of the conditions in tossing a coin, so that they have some influence on the way it falls.

On the other hand, consider the act of shooting a bullet at a target. Because the man takes aim, the ultimate result to some extent depends upon his skill. Nevertheless this act is a random process because the exact point where the bullet strikes is uncertain. The role played by the man has some effect on the ultimate result, and the role played by chance has some effect. Thus chance may be regarded as a co-agent in the final result. In fact most random processes are of the type in which chance may be regarded only as a co-agent in the ultimate result.

Statistical independence

We may classify random situations into the following two categories:

(1) experimental processes
(2) non-experimental processes.

An experimental random process is one which can be repeated for an indefinite number of instances. In each instance the same circumstances are fulfilled and one result is produced. Thus if the experimental process is repeated five times, a series of five results is produced. All other processes fall into the category of non-experimental processes.

Experimental processes are very useful in gaining insight to the methods of probability theory. For brevity, we shall replace the long expression experimental, random process by the single word *experiment*. Thus an experiment is a random process that can be repeated indefinitely, where each repetition is made under the same circumstances and produces one result. Twenty repetitions of the experiment would produce a series of twenty results. Each result, of course, must be one and only one of the possible results of the experiment. An example of an experiment is the tossing of a coin. This experiment can be repeated an indefinite number of times. For each repetition, either H or T happens. Thus a series produced by 20 repetitions might be

HHHTTHHTHTHHTTTHTHTT

The intuitive concept of experiments repeated under identical conditions leads to the notion of *statistical independence*. When a scientist says that two experiments are performed under identical conditions, he implies independence, that is, he implies that the result of either experiment has no influence on the result of the other experiment. For example, if we toss a coin twice under identical conditions, then the two tosses are independent. The appearance of heads or tails on toss 1 does not influence which side appears on toss 2.

The notion of statistical independence applies not only to repetitions of the same experiment under identical conditions, such as successive tossings of a coin, but also applies to different experiments that have nothing to do with each other. Such experiments can either be done simultaneously (i.e. at the same time) or sequentially (i.e. in a time succession). For example, suppose we toss a coin and throw a die, either simultaneously or sequentially. It seems evident that the fall of the coin has nothing to do with the fall of the die, and accordingly we say that these two experiments are independent.

Classic definition of probability

Now we want to look at random processes that have a finite number of possible results. Thus we may list all the cases for such a situation. In tossing a coin there are two cases, namely H, T. In tossing a die there are six cases, namely 1, 2, 3, 4, 5, 6. For many such finite processes, there exists a symmetry among the cases. For example, a die is in the form of a cube and is manufactured from a homogenous material. Thus its six faces have symmetry that tells us that the turning up of one side should not be more propitious than the turning up of any other side. We must use our judgement to decide for each particular process whether its cases do or do not have the necessary kind of symmetry. In other words, for the process in question we must decide whether or not any case is more propitiously endowed than any other case. To make this decision we must make use of our accumulated experience as well as theoretical reasoning by analogy.

When the cases of a random process do have this kind of symmetry, that is, when no case is more propitious than any other case, then each case is *equally likely.* An *equally-likely case* is called a *chance.* We may then define the *probability* of an event as the ratio of the number of favorable chances to the total number of chances. In symbols we write $P(A)$ to stand for the probability of event A. Then the definition is

$$P(A) = \frac{\text{Number of favorable chances}}{\text{Total number of chances}}$$

This definition is called the *classic definition of probability.* For example, when we toss a good coin, H has 1 chance in 2, so the probability of H is ½. When we throw a good die, the event 5 (i.e. the face 5 lands up) has 1 chance in 6, so its probability is 1/6. When we say that the event of each face on a good die has a probability of 1/6, we always must take into consideration that a so-called "good" die can only be realized in practice to a certain approximation. The same situation holds for all the questions studied by means of probability theory.

The difference between a mathematical model and a mathematical theory should be noted. A mathematical theory does not need any connection whatsoever with the real world. A mathematical model is a mathematical theory that can be applied to real phenomena. The definition of probability given here represents a mathematical model. There are many applications of this model to real phenomena. In fact this probability model was discovered in the Renaissance by the empirical study of some real phenomena, namely the tossing of dice and the dealing of cards in games of chance.

Relative frequency definition of probability

Let us now see how we can interpret a probability, as say the probability of 1/6 that the face 5 lands up in the toss of a die. We make use of the so-called *Law of Large Numbers,* which states:

> If an experiment is repeated more and more times, then the relative frequency of a given event tends closer and closer to the probability of the event.

If the die is cast again and again, then the relative frequency

$$f = \frac{\text{Number of times face 5 lands up}}{\text{Total number of tosses}}$$

will fluctuate, but it comes closer and closer to the value 1/6, namely the probability of the face 5 landing up. As a result we can interpret the probability of an event as the proportion of the times it will occur in the long run. In saying that the probability is 1/6, we mean that if the experiment is repeated a great many times the event will occur about 16 2/3 percent of the time. Note that we do not say that the event must occur 1 time out of 6, or 10 times out of 60, or 100 times out of 600. We only say that the relative frequency of the event in 600 tries will generally be closer to the true probability 1/6 then the relative frequency of the event in 60 tries. In turn, the relative frequency in 60 tries will be a better estimate of the probability 1/6 than the relative frequency in only 6 times. In short, things take time.

As a consequence of the statistical regularity indicated by the law of large numbers we now have a way of estimating the values of probabilities. Many events cannot be put into the framework of the classical definition of probability, which requires the existence of equally likely cases. For example, let us observe a number of automobiles of a given year and model, and suppose we want to predict whether each automobile five year hence is still in operation. Let us suppose that with expert mechanics we are able to make detailed examinations of each car, and also are able to collect pertinent information as to the occupation and habits of the owners. However it is impossible to make exact predictions with regard to the outcome of one particular automobile, since the causes leading to the ultimate result are far too numerous and too complicated to allow any precise calculation. We can think of the five year time-span of each car as an experiment, and the experiment is a success if the car is still in operation after five years. As we have said we are not able to predict individual results, but as soon as we turn our attention from the individual cars to the whole sequence of cars we find that the relative frequency of successes shows a striking regularity as we include more and more cars. More specifically, the relative frequency for more and more cars shows a marked tendency to become more and more constant. If such a series of cars could be indefinitely continued under uniform conditions, then we would expect the relative frequency to approach some definite ideal value which we call the probability. It is a common experience to mankind that this stability of a relative frequency usually appears in long series of repeated experiments performed under uniform conditions. Moreover in situations where this statement is not true, a careful examination will usually disclose some definite lack of uniformity. However this statement represents a conjecture that can neither be proved nor disproved either mathematically or empirically, and all we can say is that its validity is supported by our experiences in life. The resulting meaning assigned to the word *probability* is called the *frequency interpretation of probability*. We can, therefore, define the probability of an event as the limiting ratio of the number of favorable results to the total number of results, as the total number of repetitions of the random experiment becomes larger and larger. The definition can be written as

$$P(A) = \frac{\text{Number of favorable results}}{\text{Total number of results}}$$

where it is understood that the total number of repetitions gets very large. Because this ratio is the relative frequency of the event A, we call this definition the *relative frequency definition of probability*.

In real life there is an upper bound to the number of times we can perform an experiment, so we can never be sure that we have obtained a large enough total to insure the accuracy of our probability estimate. In such cases we must do the best we can. For example, if we could only study the history of 1000 automobiles, and found that 650 were still in operation after 5 years, we would assign the relative frequency 650/1000 = 0.65 as the probability of success.

Personal, or subjective, definition of probability

There is a third definition of probability, which is often useful in the case of a single non-repetitive event. The *personal, or subjective, definition of probability* says that the probability is the measure of a person's belief as to the occurrence of the event. Such personal probabilities are arrived at by subjective reasoning that is difficult or impossible to pin down to a formula or method as we were able to do in the case of the other two definitions. However the concept is useful, especially in dealing with situations where there is little or no direct evidence. Sometimes such probabilities are referred to as intuition or guesses. For example, a scientist might subjectively arrive at the figure 0.75, or 3 chances out of 4, as the probability of success in a new venture in which neither he nor any one else has had any previous experience. A mathematician might assign a probability of 99 percent that Fermat's last theorem will never be proved.

Events

As we have seen the universe of a random process is made up of all the possible results, or cases. These cases are elemental (i.e., part of a case is not admitted), exhaustive (i.e., one case happens) and exclusive (i.e., only one case happens). In brief, whenever the circumstances of a random process are fulfilled, one and only one case happens. For example, in the tossing of a die there are six cases: 1, 2, 3, 4, 5, 6. An *event* is a set of cases. In our die example, the event *odd* is the set consisting of the three cases 1, 3, 5. The cases that make up an event are called the members of the event. We say that an event happens when and only when one of its members happens.

The *universal event* is the event made up of all the cases. It is the universe of the random process under consideration. Because the universal event always happens, it is a sure event and we assign it a probability of one. The *empty* (or *null*) *event* is the event made up of none of the cases. Because the null event never happens, we assign it a probability of zero. The probabilities of all the events must lie between 0 and 1.

For every event there is a contrary event. If we let A designate an event, we shall let $\bar{A}$ designate the *contrary event*. If we remove the members of A from the universe, then we are left with the contrary event $\bar{A}$. The two events A and $\bar{A}$ have no members in common, but everything in the universe is either a member of A or a member of $\bar{A}$. For example, in the tossing of a die, the event *odd* is made up of the number 1, 3, 5. The contrary event $\bar{A}$ is thus made up of 2, 4, 6, and is in fact the event *even*. If the probability of the event A is $P(A)$ then the probability of the contrary event is

$$P(\bar{A}) = 1 - P(A).$$

In other words, the sum of the probabilities of an event and the contrary event is equal to one.

Two events can have members in common. For example, in the toss of a die, the event *low* is made up of the cases 1, 2, 3, whereas the event *odd* is made up of the cases 1, 3, 5. Hence these two events have the cases 1 and 3 in common. If either 1 or 3 occurs, then both the event *low* and the event *odd* happens. Let us designate the event made up of the common cases, namely 1 and 3, by the notation

$$low \cap even$$

which we read as *low intersection even*. That is, the *common event* $A \cap B$ is defined as the event made up of the common members of the separate events A and B. These common members are cases that are members of both of the events A and B.

Two events are *incompatible* provided that they have no members in common. In other words, two events are incompatible if and only if they cannot both happen at the same time. Alternate names for incompatible events are *disjoint events* or *mutually exclusive events*. Because the common event $A \cap B$ is made up of the members common to the events A and B, we see that the events A and B are incompatible if and only if their common event $A \cap B$ is the empty event. For example, the events *odd* and *even* are incompatible. An event and its contrary event are always incompatible, that is,

$$A \cap \bar{A} = None$$

People who swim and play tennis can form a union. The union is made up of people who swim or play tennis. But suppose a person both swims and plays tennis. Certainly he will be a member of the union, but he will be counted as only one member, not as two members. Thus he has no preferential status to a person who swims but does not play tennis, who is also counted as one member of the union. Likewise, a person who plays tennis but doesn't swim is counted as one member. Suppose there are 100 people who swim (regardless of whether or not they play tennis), 70 people who play tennis (regardless of whether or not they swim) and 20 people who both swim and play tennis. Thus the union has

$$100 + 70 - 20 = 150 \text{ members.}$$

In general, given any two events A and B we can form their *union*

$$A \cup B$$

which we read as *A union B*. The event $A \cup B$ is the event whose cases are members of at least one of the events A and B.

When the events A and B are incompatible (i.e. A and B have no members in common) then the members of the union $A \cup B$ are simply all the members of A together with all the members of B. Because there is no member that both belongs to A and B, there is no possibility of double-counting such a member.

One of the most important, and also the simplest, rules used in the calculation of probabilities is the *addition rule in the case of incompatible events*. For example, on Nevada roulette there are 18 chances in 38 for red and 18 chances in 38 for black. Note the 0 and 00 on the Nevada roulette wheel are neither red nor black. The events *red*, *black* are incompatible, either *red* can happen or *black* can happen, but not both. Thus the united event *red* $\cup$ *black* has

$$18 + 18 = 36$$

chances in 38 to happen. Thus the sum of the probability that *red* happens plus the probability that *black* happens gives the probability that *red* $\cup$ *black* happens:

$$\frac{18}{38} + \frac{18}{38} = \frac{36}{38}.$$

We may summarize our reasoning as follows:

If we verify that two events A, B are incompatible $A \cap B = 0$, then we find the probability of their united event $A \cup B$ by adding their separate probabilities:

$$P(A \cup B) = P(A) + P(B).$$

More will be spent on the addition rule in the case of incompatible events in the next chapter.

Expectation of an event

The probability of an event is not the only quantity in which we are interested in the study of a random process. For example, it is very different to have 1 chance in 10 to gain $100 or to have 1 chance in 10 to gain $100,000. This consideration leads to the notion of *mathematical expectation*, or to use a shorter term *expectation*, of an event.

If a person has 1 chance in 10 to gain $100, we say that his expectation is

$$\$100 \ (1/10) = \$10.$$

If he has 1 chance in 10 to gain $100,000, his expectation is

$$\$100,000 \ (1/10) = \$10,000.$$

If a person has 1 chance in 2 to gain $200, his expectation is

$$\$200 \ (1/2) = \$100.$$

Suppose that the only prize given at a lottery is $100,000, and that there are 200,000 tickets sold at $1 each. The expectation of a ticket is then

$$\$100,000 \ (1/200,000) = \$0.50$$

even though the cost of the ticket was $1. Examining this situation, we see that the people who organized this lottery realize $200,000 from ticket sales, and only pay out $100,000 as a prize, thereby making a profit of $100,000. On the other hand each of the 200,000 ticket holders paid $1 for a ticket whose expectation was only $0.50.

The *expectation* of a contingent gain is *the product of the gain times the probability of realizing this gain*. We can say that the expectation is the value of a gain whose attainment is not certain.

When it is a matter of a loss instead of a gain, we shall consider a loss to be a negative gain. For example, if a person has 1 chance in 50 to lose $100, his expectation is

$$-\$100 \ (1/50) = -\$2.$$

The expectation is negative. A negative expectation is the value of a loss whose realization is not certain but only contingent.

If there is a probability of 1/20 to sustain a loss of $800, the expectation is

$$-\$800 \ (1/20) = -\$40.$$

In order to avoid this risk of losing $800, we may say that $40 is the "fair" amount that should be paid.

Suppose that for a toss of a coin, a player receives $100 if the coin lands *H* (where as before *H* stands for heads). The probability of *H* is 0.5. We say that the expectation of the player is

$$\$100 \ (0.5) = \$50.$$

Suppose that for two consecutive tosses of a coin, another player receives $100 if both tosses result in *H*. The probability that both tosses land *H* is 0.25. The expectation of the player is then

$$\$100 \ (0.25) = \$25.$$

The expectation of a player is the product of his possible gain times the probability that he has of realizing this possible gain.

Summing up, we have: *The expectation of an event is the product of the gain received if the event happens times the probability that the event happens.*

The expectation is a fictitious or imaginary sum of money. It does not ordinarily correspond to a possible value of gain or of loss. If a person has 1 chance in 10 to gain $40 his expectation is $4. But $4 is not a possible value of his gain. The only possible values that can happen are $0 and $40. If a person has 1 chance in 5 to lose $10 his expectation is a loss of $2. Nevertheless, either a loss of $0 or a loss of $10 will happen, never a loss of $2.

Expectation of several events

An advantage of the notion of expectation is the following. The combination of different probabilities can be complicated, but the combination of different expectations is simple and intuitive. It is easy to see that expectations can be added together as sums of ordinary money.

The rule is: *The expectation for several events is the sum of the expectations for each of the events.*

This property of addition can make the calculation of expectations very easy in many applications. Suppose a coin is tossed 3 times. If it lands H on toss 1 a player receives $10. If it lands H on toss 2 he receives $20. If it lands H on toss 3 he receives $30. Thus we have

Event	Gain	Probability	Expectation
H on toss 1	$10	0.5	$5
H on toss 2	$20	0.5	$10
H on toss 3	$30	0.5	$15

Therefore his total expectation is

$$\$5 + \$10 + \$15 = \$30$$

As we have seen, a negative gain is a loss. In dealing with expectations, we must be prepared to consider negative gains (or losses) as well as positive gains.

Suppose a coin is tossed. A player receives $100 if the coin lands H and loses $100 if the coin lands T. What is his expectation? The loss of $100 is the same as a gain of −$100. Hence we have the table:

Event	Gain	Probability	Expectation
H	$100	0.5	$50
T	−$100	0.5	−$50

Hence his total expectation is

$$\$50 - \$50 = 0.$$

Advantageous, equitable, and disadvantageous games

A game is *advantageous* to a player if his expectation is positive. It is *disadvantageous* if his expectation is negative. It is neither advantageous nor disadvantageous when his expectation is zero. We then say that the game is *equitable*.

For example, suppose a player has 1 chance in 3 to gain $24 and 2 chances in 3 to lose $12. His expectation is

$$\$24 \ (1/3) - \$12 \ (2/3) = 0$$

This game is equitable.

If a game is made up of several trials, the expectation of the game is the sum of the expectations of the various trials that compose the game. It is understood that all the trials are necessarily played.

In particular, if all the trials are identical, the expectation of the game (composed of n trials) is equal to the product of n times the expectation of one trial. Thus if each trial is equitable, the game is equitable. As a result there is no way to make advantageous or disadvantageous a game of which each trial is equitable.

Betting on events

Gambling is the act of betting on unsure events. Life and people's ingenuity provide ample events upon which people can bet.

To place a bet one must first know the odds given for the event. The *bettor's odds* consist of two numbers separated by the word *to,* such as

$$8 \; to \; 5.$$

The bettor's odds represent the ratio of his potential gain to his potential loss, that is

$$odds \; = \; gain \; to \; loss.$$

For example, suppose his odds are 8 to 5 for the event, and the better bets $5 on the event. If the event happens he wins. His winnings are $13, made up of his $8 gain plus his $5 bet. If the event doesn't happen he loses. His loss is $5, made up of his $5 bet.

Another way of thinking about the bettor's odds is that the odds represent *his potential gain for his bet.* For example, suppose the odds are 6 for 1 for an event, and a bettor bets $10 on the event. If the event happens, his gain is $60 for his $10 bet, making his total purse equal to $70. If the event doesn't happen, his loss is his bet of $10.

Application to life insurance

According to the mortality tables of an insurance company, the probability that a 25 year old man will live at least one more year is 0.993, whereas the probability that he will die within the year is 0.007. A $10000 1-year term life insurance policy is an agreement whereby the insurance company will pay to the man's survivors $10000 if the man dies within the year, but pays nothing if the man lives for the entire year. The company offers such a policy to a 25-year old man for a premium (that is, cost to the man) of $100.

The company faces the following situation:

Event	Gain to Company	Probability	Expectation
Man lives	$100	0.993	$99.30
Man dies	-$9900	0.007	-$69.30

Hence the expectation of the company is $99.30 minus $69.30, which is $30.00 from which the company must pay costs, taxes, and dividends. If this expectation were negative, the company could not stay in business. As a game, life insurance is advantageous to the company and disadvantageous to the policy holder.

Exercises

1. Games may be divided into three categories: (1) pure chance, (2) chance and skill, (3) pure skill. Into which categories would you place chess, dominos, most card games, roulette, dice?

2. It is generally impossible for a card player to have in his mind, in a complete way, all of the different probabilities. One cannot fully comprehend the prodigious number of different possible results that can be obtained from a pack of cards. The number of possible arrangements of 52 cards is expressed by an 8 followed by 67 other digits. Since the player is not able to completely know all the probabilities, can he appeal to his experience and intuition to supplement his partial knowledge, so that his skill and cleverness play a part, and the game no longer appears as completely dependent upon chance?

3. In a game of coin tossing, dice, or roulette a player can have in his mind the probabilities of the different events which he is faced with. Can we say that a knowledgeable player is one who knows these probabilities and acts on their basis, whereas an unknowledgeable player is one who does not know the probabilities and is under the illusion that his skill and cleverness play a part?

4. A has 3 silver dollars and B has 2 silver dollars. The coins are all tossed, under the agreement that the player having the greatest number of heads shall win all the 5 silver dollars, but in case of a tie B shall win. What is the expectation of A?

5. In the above problem, what is the expectation of player A if the rules are changed so that the 2 players agree to begin again in case of a tie?

6. The French roulette wheel is used in Monte Carlo. It has 37 numbers: 0, 1, 2, 3, 4, . . . , 34, 35, 36. The casino pays odds of 35 to 1 for single-number events. What is the expected loss of a bet of 100 francs played on a single number? *Ans.* 2.70

7. The event *odd* on the French roulette wheel is made up of the 18 numbers: 1, 3, 5, . . . , 33, 35, whereas the event *even* is made up of the 18 numbers: 2, 4, 6, . . . , 34, 36. If you bet 100 francs on *odd* then (1) if *odd* happens, your winnings are a gain of 100 francs plus your bet of 100 francs, (2) if *even* happens, you lose your 100 francs bet, and (3) if 0 happens, your bet is left where it stands and the wheel spun again. On this second spin: (1) If *odd* happens you take your bet of 100 francs back, (2) if *even* happens you lose your 100 francs, and (3) if 0 happens, the wheel is spun a third time, and so on. What is your expected loss? *Ans.* 1.35

8. Two dice are thrown. What is the probability that they show 7 or 11?

CHAPTER 4

RULES FOR PROBABILITIES

Addition rule for probabilities

The fundamental rule upon which the rule for the addition of probabilities rests is the addition rule. The *addition rule* is:

If one thing can be done in *m* different ways and another thing can be done in *n other* different ways, then either one thing or the other can be done in the sum of *m* plus *n* different ways.

This rule is basic, and goes back to when you first learned how to add. For example, if we put 5 apples and 6 oranges into a box, then the number of apples and oranges in the box is 5 + 6, or 11.

As another example, if John knows 4 kinds of trees and Mary knows 3 other kinds, then John or Mary know 4 + 3 = 7 different kinds. Suppose John knows oak, elm, maple and birch, and Mary knows cedar, pine, and spruce. Then oak, elm, maple, birch, cedar, pine and spruce are known by John or Mary. However, if John knows 4 kinds of trees and Jane knows 3 kinds, then we cannot conclude that John or Jane know 4 + 3 or 7 different kinds. The reason is that Jane might not know 3 *other* kinds, but instead knows some of the same kinds as John. Suppose that Jane knows oak, elm and pine. Then oak, elm, maple, birch, and pine are known by John or Jane, which are 5 different kinds.

Thus the word *other* is an essential part of the addition rule.

A set that has only a finite number of members is called a finite set. Let this finite number be denoted by *n*. The number *n* is a positive whole number, such as 1, or 2, or 5, or 5000, or 500,000. We exclude $n = 0$, which would correspond to the empty set. Instead of saying

a set with n numbers

it is much more convenient to use a shorter expression. Hence let us agree that the expression

n-set

means "a set with *n* members."

Now we can state the *addition rule* in terms of sets.

Let *A* be a *m*-set and *B* be a *n*-set. Let *A* and *B* have no members in common, that is, let the common set satisfy

$$A \cap B = None.$$

Then the united set

$$A \cup B$$

is an $(m + n)$-set.

Let us now make use of the classic definition of probability. We deal with a random process that has these two properties:
(1) The process has only a finite number of cases
(2) No case is more propitously endowed than any other case.
The *classic definition* is:

The probability of an event is the ratio of the number of favorable cases to the total number of cases.

Suppose we have an urn that holds 20 like balls, among which:

5 are colored amber (A)
7 are colored blue (B)
8 are other colors.

Let the desired event be the event of extracting an amber or blue ball. This event is denoted by

$$A \cup B$$

There are 2 kinds of favorable cases for this event. One kind is represented by the amber balls, that is, the cases of the event A. The other kind is represented by the blue balls; that is, the cases of the event B. The event A has no cases in common with the event B, because if the extracted ball is amber then it cannot be blue, and if the extracted ball is blue then it cannot be amber. Thus the two events A, B are incompatible; that is $A \cap B = None$. The event A has 5 cases in 20. The event B has 7 cases in 20. The 5 cases of A are distinct from the 7 cases of B. These two events cannot both happen at the same time. Because the events A, B have no cases in common, the number of cases for the event $A \cup B$ is the sum of the cases for A plus the cases for B. That is, the number of cases for the event $A \cup B$ is Because the cases add, the probabilities also add. Hence the probability of the event $A \cup B$ is the sum of the probability of A plus the probability of B. That is, the probability of the event $A \cup B$ is

$$5/20 + 7/20 = 12/20.$$

This equation illustrates the *rule for the addition of probabilities:*

$$P(A) + P(B) = P(A \cup B) \quad \text{provided } A \cap B = None.$$

For ease of memory, it may be stated as follows:

If the events A, B are incompatible, then the probability of the united event $A \cup B$ is equal to the sum of the probability of A plus the probability of B.

The rule states, simply, that the probabilities of two incompatible events can be added. The sum of the probabilities is the probability that one *or* the other of these two incompatible events happens.

In the application of this rule, it is very important to verify that the condition of incompatibility is fulfilled. Suppose that there are 5 air routes from London to New York and 3 air routes from London to Moscow. These sets of air routes are incompatible. Thus the number of air routes going from London to either New York or Moscow is

$$5 + 3 = 8.$$

Suppose that there are 30 planes at the London airport, and each plane is going to fly a different air route. You haphazardly get on one of the airplanes. Then you have 5 chances in 30 to go to New York, and 3 chances in 30 to go to Moscow. Thus you have 8 chances in 30 to go to either New York or Moscow, and so the probability of this event is 8/30.

Now let us suppose that 1 of the 5 air routes from London to New York includes a stop at Iceland. Also suppose there is no other air route from London to Iceland. Although there are 5 air routes to New York and 1 air route to Iceland, the number of air routes to either New York or Iceland is not 6, but 5. In other words, the air route to Iceland is the same as one of the air routes to New York, and so the event of going to Iceland is not incompatible with the event of going to New York. Your probability of going to Iceland or New York is 5/30, not 6/30.

Problem: Three coins are tossed. What is the probability of *at least* 1 *T?* Note: T = tails and H = heads.

Solution: The tossing of 3 coins has 8 cases in total, namely

HHH, HHT, HTH, THH, HTT, THT, TTH, TTT.

The event 1 T (i.e. 1 tails shows) has 3 cases, namely

HHT, HTH, THH.

The event 2 T (i.e. 2 tails show) has 3 cases, namely

HTT, THT, TTH.

The event 3 T (i.e. 3 tails show) has 1 case, namely

TTT.

The event of *at least 1 T* is the union of the events 1 T, 2 T, 3 T, that is

at least 1 $T = 1\ T \cup 2\ T \cup 3\ T.$

The events 1 T, 2 T, 3 T are incompatible. Hence the event of *at least* 1 T has

$$3 + 3 + 1 = 7$$

chances, so its probability is 7/8. Because

$$P(1T) = 3/8$$
$$P(2T) = 3/8$$
$$P(3T) = 1/8$$

we see that

$$\frac{7}{8} = \frac{3}{8} + \frac{3}{8} + \frac{1}{8}$$

or

$$P(1T \cup 2T \cup 3T) = P(1T) + P(2T) + P(3T).$$

The equation represents the *addition rule for probability for three incompatible events.*

   Problem: Ann and Bob toss a coin under the following conditions. If toss 1 shows H, Ann wins. If toss 1 shows T, however, the coin must be tossed 2 more times. Then, if out of the 3 tosses, H shows at least 2 times, Ann also wins. What is the probability that Ann wins?

   Solution: We could reason as follows. Ann wins in 2 different ways. The first way is for H to come up on toss 1. The probability of this event is 1/2. The second way is for H to come up at least 2 times out of 3 tosses. The probability of this event is also 1/2. Thus the probability that Ann wins is the sum of these 2 probabilities. This sum is

$$\frac{1}{2} + \frac{1}{2} = 1$$

which says that it is certain that Ann wins. But this is absurd. We must consider the set of all possible outcomes, namely

$$HHH, \ HHT, \ HTH, \ THH, \ HTT, \ THT, \ TTH, \ TTT.$$

Now the cases favorable to Ann winning by the first way (namely, by H coming up on toss 1) are

$$HHH, \ HHT, \ HTH, \ HTT$$

so indeed the probability is 4/8, or 1/2, for her winning by the first way. Also the cases favorable to Ann winning by the second way (namely, by H coming up in 2 out of three tosses) are

$$HHH, \ HHT, \ HTH, \ THH$$

so the probability is 4/8, or 1/2, for her winning by the second way. But for 3 of these cases, namely

$$HHH, \ HHT, \ HTH,$$

Ann wins directly by the first way (namely, H on toss 1), and so toss 2 and 3 need not be made. Thus the two ways that Ann might win are not incompatible, so it was not legitimate to add their probabilities. In other words, the chances

$$HHH, \ HHT, \ HTH$$

are common to both ways. The chances favorable to Ann winning one way or the other are

First way

THH, HHH, HHT, HTH, HTT

Second way

so her chances are 5 in 8. Thus the probability of Ann winning is 5/8.

Multiplication rule for probabilities

Another general rule, which like the addition rule is extremely useful, is the *multiplication rule*. It is

If one thing can be done in M different ways, and *then* another thing can be done in N different ways, it follows that both things can be done in the product of M times N different ways.

Another statement of the *multiplication rule* is

If there is an M-way choice for thing 1 and *then* there is an N-way choice for thing 2, it follows that there is an MN-way choice for both things.

This rule is basic. It was first taught when you learned how to multiply numbers.

For example, if you first choose one of coffee, tea, or milk and then either a ham sandwich or a peanut butter sandwich, there are $(3 \times 2) = 6$ different choices that you have, namely

coffee, ham sandwich   coffee, peanut butter sandwich
tea, ham sandwich   tea, peanut butter sandwich
milk, ham sandwich   milk, peanut butter sandwich

Suppose there are 3 trails to the top of Mount Washington. Thus there are 3 ways up and 3 ways down, so that there are 3 times 3, or 9 ways of making one round trip. If the trails are called A, B, C then these 9 round trips are

$$AA \quad AB \quad AC$$
$$BA \quad BB \quad BC$$
$$CA \quad CB \quad CC.$$

Suppose now that we don't want to come down by the same trail as the one we went up. Now there are 3 trails to go up, but only 2 trails that we can come down on. Hence the number of round trips is 3 times 2, or 6. They are

$$AB \quad AC$$
$$BA \quad BC$$
$$CA \quad CB.$$

Problem: Two persons get into an airplane where there are 9 vacant seats. In how many different ways can they seat themselves?

Solution: Person 1 can take any of the 9 vacant seats. Thereafter person 2 can take any of the 8 seats that are left. Hence there are $9 \cdot 8$, or 72 different ways that they can take their seats.

The multiplication rule is very useful in computing probabilities. In the problems here we make use of the fundamental classic definition:

$$\text{Probability} = \frac{\text{Number of favorable cases}}{\text{Number of all cases}}$$

Problem: Two cards are drawn from a well-shuffled pack of 52 cards. What is the probability that both cards drawn are kings?

Solution: Since there are 52 cards in the pack, the first card can be drawn 52 ways. After the first card has been withdrawn, there are 51 cards remaining in the pack, so the second card can be drawn in 51 ways. Therefore, the total number of ways to draw 2 cards is

$$52 \cdot 51 = 2652$$

which is the number of all the cases. To find the number of cases favorable to the event of drawing two kings, we observe that there are 4 kings in the pack. Thus the first king can be drawn in 4 ways. After the first king has been drawn, there are 3 kings left, so the second king can be drawn in 3 ways. Therefore the number of favorable cases is

$$4 \cdot 3 = 12$$

Hence there are 12 cases in 2652 that both cards drawn are kings, so the required probability is

$$P(2K) = \frac{4 \cdot 3}{52 \cdot 51} = \frac{12}{2652} = \frac{1}{13 \cdot 17} = \frac{1}{221}.$$

That is, there is 1 chance in 221 of drawing 2 kings if the first card is not returned to the pack before drawing the second card.

Problem: Two cards are drawn from a pack of 52 cards, the first card being returned to the pack before the second card is drawn. (By this description we mean: The cards are shuffled and the first card is drawn and noted. This card is then returned to the pack, the pack is shuffled again, and the second card is drawn and noted.) What is the probability that both cards drawn are kings?

Solution: There are 52 ways of drawing the first card. There are also 52 ways of drawing the second card, because by returning the first card drawn, the pack is restored to its original number of 52 cards. Thus the number of ways to draw both cards is $52 \cdot 52$. There are 4 ways of drawing the first king, and because the pack is restored, there are also 4 ways of drawing the second king. Thus the number of ways to draw both kings is $4 \cdot 4$. Hence the required probability is

$$P(2K) = \frac{4 \cdot 4}{52 \cdot 52} = \frac{1}{13 \cdot 13} = \frac{1}{169}$$

That is, there is 1 chance in 169 of drawing 2 kings if the first card drawn is returned to the pack before drawing the second card.

Problem: 3 chests, identical in appearance, each have 2 drawers. Chest 1 has a G (Gold) coin in each of drawers 1 and 2. Chest 2 has a S (Silver) coin in each of drawers 1 and 2. Chest 3 has a S (Silver) coin in drawer 1 and a G (Gold) coin in drawer 2. See the figure below

	Chest 1:	Chest 2:	Chest 3:
Drawer 1	G	S	S
Drawer 2	G	S	G

A chest is chosen at random. What is the probability that its 2 coins are of different metals.

Solution: Since outwardly the chests are indistinguishable from each other, we recognize each chest as having 1 chance to be drawn. Among the 3 chests, only 1 chest has coins of different metals, namely chest 3. Therefore the required probability is 1/3.

Problem: Let there be 3 chests as in the foregoing problem. A chest is chosen at random, one of its drawers is opened, and a S (silver) coin is found. What is the probability that the other drawer contains a G (gold) coin?

Solution: The fact that S was found in one drawer leaves only 2 possibilities as to the content of the other drawer, namely S, G. Hence we might be tempted to reason that the probability of G in the other drawer is 1/2. Nevertheless, this reasoning is false, because each of the possibilities cannot be regarded as having 1 chance each. Before the chest is chosen and one drawer opened, there are 6 possible results, or cases,

Drawer 1 of chest 1   Drawer 1 of chest 2   Drawer 1 of chest 3
Drawer 2 of chest 1   Drawer 2 of chest 2   Drawer 2 of chest 3

each one of the 6 representing 1 chance. But as soon as S is found in the drawer that is opened, the 3 cases

Drawer 1 of chest 1
Drawer 2 of chest 1   Drawer 2 of chest 3

became impossible, and so there remain only 3 cases, namely

Drawer 1 of chest 2
Drawer 2 of chest 2   Drawer 1 of chest 3

Each of these cases represents 1 chance. In these 3 chances, there is 1 chance, namely Drawer 1 of chest 3, that the other drawer contains G. There the required probability is 1/3.

We may extend the *multiplication rule* to doing 3 things, one after another. Thus, we have:

If thing 1 can be done in N_1 different ways, and if *then* thing 2 can be done in N_2 different ways, and if *then* thing 3 can be done in N_3 different ways, it follows that all the things can be done in $N_1 N_2 N_3$ different ways.

Problem: 10 persons compete for 3 prizes. In how many different ways can the prizes be awarded?

Solution: Prize 1 can be given in 10 different ways. Thereafter prize 2 can be given in 9 different ways. Thereafter prize 3 can be given in 8 different ways. Hence there are $(10)(9)(8) = 720$ different ways to give the prizes.

The *multiplication rule* can be extended to doing any number of things.

If series of things can be done successively as follows:

Thing 1 in N_1 different ways,
then thing 2 in N_2 different ways,
then thing 3 in N_3 different ways,

. . .

then thing K in N_K different ways,

then all the things can be done in
$$N_1 N_2 N_3 \ldots N_K$$

different ways.

Problem: The probability that a newspaper reader will read an advertisement is 0.3 and if he reads the advertisement the probability that he will buy the product advertised is 0.02. What is the probability that he will read the advertisement and then buy the product?

Solution: The tree diagram is shown in the figure below

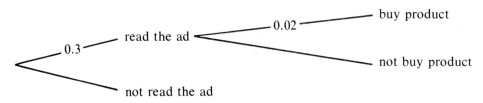

The required probability is

$$(0.3)(0.02) = 0.006.$$

Problem: There are 5 coins in purse 1, 4 of which are S (silver) and 1 G (gold). There are 5 coins in purse 2, all of which are S (silver). Suppose that 4 coins are drawn at random from purse 1 and put into purse 2. Then 4 coins are drawn at random from purse 2 and put into purse 1. A person may now pick whichever purse he pleases. Which purse should he pick?

Solution: At the end each purse contains the same number of coins, so he ought to pick the purse that has the greater probability to contain G (the gold coin). Now G can only be in purse 2 provided the following path is taken:

$$\frac{p_1}{}\quad \begin{array}{c} G \text{ was among the} \\ 4 \text{ coins drawn} \\ \text{from purse 1} \end{array} \qquad \frac{q_2}{}\quad \begin{array}{c} G \text{ was not among} \\ \text{the 4 coins drawn} \\ \text{from purse 2.} \end{array}$$

Instead of directly computing the probability of the event that G was among the 4 coins drawn from purse 1, let us first compute the probability of the contrary event. The contrary event is made up of one path, namely

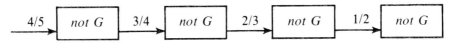

which has probability

$$q_1 = \frac{4}{5}\,\frac{3}{4}\,\frac{2}{3}\,\frac{1}{2} = \frac{1}{5}.$$

Therefore the probability of the event that G was among the 4 coins drawn from purse 1 is

$$p_1 = 1 - q_1 = 1 - \frac{1}{5} = \frac{4}{5}$$

Given that G was among the 4 coins put into purse (so purse 2 now contains 8 S and 1 G), the event that G was not among the 4 coins drawn from purse 2 is made up of one path, namely

8/9	not G	7/8	not G	6/7	not G	5/6	not G

which has the probability

$$q_2 = \frac{8}{9}\,\frac{7}{8}\,\frac{6}{7}\,\frac{5}{6} = \frac{5}{9}.$$

Hence the probability that G is in purse 2 is

$$p_1 q_2 = \frac{4}{5}\,\frac{5}{9} = \frac{4}{9}$$

so the person should pick purse 1.

Tree diagrams

The multiplication rule can be displayed graphically by means of a tree diagram.

Problem: A man has 2 suits: gray and brown and 3 neckties: red, blue and green. What choice of suit and necktie does he have?

Solution: His choice is 2 times 3, or 6. To illustrate the use of a tree diagram, we let the left fork represent the choice of a suit. There are 2 suits so the left fork has 2 branches. Thus this fork is a 2-way fork. Each of the 2 branches on the left fork lead to a separate right fork. Each of the 2 right forks represent the choice of a necktie. There are three neckties so each right fork has 3 branches. Thus each right fork is a 3-way fork. A choice of both suit and necktie correspond to a choice of path from the left to right. As we see in the figure below there are 2 · 3 = 6 such paths.

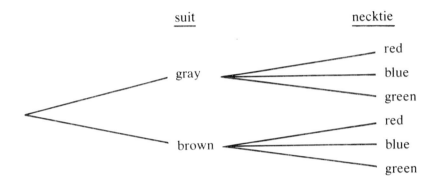

Problem: Purse 1 contains 1 dime and 2 nickels. Purse 2 contains 2 dimes and 1 nickel. You happen to take a purse and a coin from it. What is the probability that the coin is a dime?

Solution: The first process is the taking of a purse; the second process is the taking of a coin from it. Both of these processes make up the overall process illustrated in the figure below. From the figure we see there are 6 paths, each of which may be considered to represent 1 chance. Of these 6 chances, 3 result in a dime. Hence there are 3 chances in 6, or a probability equal to 3/6 = 1/2, that the coin is a dime.

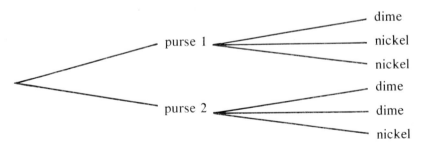

Alternative solution: Instead of constructing a tree diagram so that each path represent 1 chance, we can construct a tree diagram as shown in the figure below:

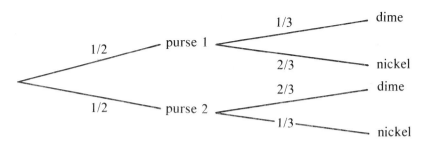

In this tree diagram, we have assigned a probability to each branch. For the first process, each purse has 1 chance in 2 to be taken, so we have assigned a probability of 1/2 to each branch. For the second process we must consider 2 situations, namely

          (1) purse 1 was taken
          (2) purse 2 was taken.

Given that purse 1 was taken, there is 1 chance in 3 for a dime, so the probability 1/3 is assigned to the *dime* branch leading out of purse 1. Given that purse 1 was taken, there are 2 chances in 3 for a nickel. So the probability 2/3 is assigned to the *nickel* branch leading out of purse 1.

Given that purse 2 was taken, there are 2 chances in 3 for a dime, so the probability 2/3 is assigned to the *dime* branch leading out of purse 2. Given that purse 2 was taken, there is 1 chance in 3 for a nickel, so the probability 1/3 is assigned to the *nickel* branch leading out of purse 2.

Let us consider the sequence of the two events

          (1) purse 1 is taken
          (2) a dime is taken from it.

This sequence of two events, which is the top path in the above figure, may be denoted by

purse 1 *then dime.*

The probabilities on the branches of this path are:

(1/2) on the branch *purse* 1,
(1/3) on the branch *dime,* given *purse* 1.

Thus the probability of this path is

P (*purse* 1 *then dime*) = (1/2)(1/3) = 1/6.

In other words, the probability of the path corresponding to *purse 1* on the first process and *then dime* on the second process is equal to the product of the probabilities associated with each branch along the path. Likewise the probability of the path

purse 2 *then dime*

is

$$(1/2)(2/3) = 2/6.$$

The event of taking a dime is the union of the two events

purse 1 *then dime*,
purse 2 *then dime*,

which are incompatible. That is,

dime = (*purse* 1 *then dime*) ∪ (*purse* 2 *then dime*)

where

(*purse* 1 *then dime*) ∩ (*purse* 2 *then dime*) = *None*.

Hence by the addition rule for probabilities

P (*dime*) = P (*purse* 1 *then dime*) + P (*purse* 2 *then dime*)
= 1/6 + 2/6 = 1/2.

Exercises

1. A card is drawn from a well-shuffled pack of 52 cards. What is the probability of getting a king or a queen?

2. Two coins are tossed. What is the probability of *at least* 1 *T* (i.e. at least one tails)?

3. In how many ways can 3 letters be put into 3 addressed envelopes, one letter into each envelope?

4. A friend shows me 4 detective books and 8 western books, and lets me choose one of each. What choice have I?

5. In how many ways can we select a consonant and a vowel out of the English alphabet?

6. There are 2 red balls (labeled $R1, R2$) and 3 black balls (labeled $B1, B2, B3$) in a bag. One ball is drawn. What is the probability that it is black?

7. The contents of the bag are the same as the foregoing problem, namely $R1$, $R2, B1, B2, B3$. For this problem, however, we suppose that one ball is drawn, its color is *unnoted*, and it is laid aside. Then another ball is drawn. What is the probability that the second ball drawn is B *given* that the first ball is *unknown*? (*note:* We denote this probability by $P(B\ given?)$ which is read "*P of B given unknown*".)

8. The contents of the bag are the same as the foregoing problem, namely $R1$, $R2, B1, B2, B3$. For this problem, however, we suppose that one ball is drawn, its color is *noted*, it is *red*, and it is laid aside. Then another ball is drawn. What is the probability that the second ball is *black*, *given* that the first ball drawn is *red*? (*Note:* We denote this probability by $P(B\ given\ R)$ which is read "*P of B given R*".)

CHAPTER 5

BINOMIAL DISTRIBUTION

Family make-up

A newly married couple is planning to have two children. What is the chance of one boy and one girl in either order? Let us assume for the moment that boys and girls are born with equal probability, namely 0.5. Each child is as likely to be a girl as a boy. The two children may then be two girls, or girl and boy, or two boys. These three possibilities, however, are not equally likely. What we really have is four equally-likely possibilities, which are girl-girl, girl-boy, boy-girl, boy-boy. Because two of these four possibilities consist of a girl and a boy, there are two cases out of four that the couple have a mixed pair in two children, which is a probability of 0.5.

Let us now be more exact. From empirical knowledge, the probability of the birth of a boy is 0.51 and the probability of the birth of a girl is 0.49.

The first birth can result in either boy or girl, with the probabilities 0.51 or 0.49 respectively. This result may be presented as follows:

G	0.49
B	0.51

where G stands for girl and B stands for boy. Regardless of the outcome of the first birth, the second birth must result in girl or boy again with the probabilities 0.49 and 0.51 respectively, since the outcome of the first birth has no effect on that of the second. That one birth has no effect on any other birth is the notion of *independence* which is basic to the development we give here. Combining the possible outcomes of the second birth to those of the first birth we obtain:

GG	(0.49)(0.49)
GB	(0.49)(0.51)
BG	(0.51)(0.49)
BB	(0.51)(0.51)

Here GG stands for girl in the first birth and girl in the second birth and (0.49)(0.49) is the probability of this event. Here we have used the rule of the multiplication of probabilities. Likewise GB stands for girl-boy and (0.49)(0.51) is the probability of this event. The remaining two entries are similarly interpreted. The above table shows that the combination of two births leads to four possible outcomes. However, we can combine the two outcomes GB and BG because they have the common property that each contains one girl and one boy. The outcomes

GB and BG are identical if one does not consider the order in which G and B are obtained. Moreover we see that each of the events GB and BG has the same probability, namely $(0.51)(0.49)$. From this point of view only three separate outcomes are possible: (1) neither birth results in a boy, (2) only one birth results in a boy, or (3) both births result in boys. We can summarize this new way of looking at the results of two births as follows:

0B	$(0.49)^2$
1B	$2(0.51)(0.49)$
2B	$(0.51)^2$

Here 0B stands for no boys and two girls and $(0.49)^2$ is the probability of this event. Likewise 1B stands for one boy and one girl and $2(0.51)(0.49)$ is the probability of this event. This probability has resulted from use of the addition rule, for we added the probability $(0.49)(0.51)$ of GB and the probability of $(0.51)(0.49)$ of BG to obtain $2(0.51)(0.49)$. Finally, 2B stands for two boys and no girls and $(0.51)^2$ is the probability of this event.

Let us go on now and consider a third birth. Considering all possible outcomes and their probabilities we obtain the table:

GGG	$(0.49)^3$
GGB	$(0.51)(0.49)^2$
GBG	$(0.51)(0.49)^2$
GBB	$(0.51)^2(0.49)$
BGG	$(0.51)(0.49)^2$
BGB	$(0.51)^2(0.49)$
BBG	$(0.51)^2(0.49)$
BBB	$(0.51)^3$

The above columns show that there are eight cases. However we may bring together all the cases leading to the same family make-up and so obtain the table:

0B	$(0.49)^3$
1B	$3(0.51)(0.49)^2$
2B	$3(0.51)^2(0.49)$
3B	$(0.51)^3$

For example the event 2B is made up of the three cases GBB, BGB, and BBG each of which has a probability $(0.51)^2(0.49)$. Thus the probability of the event 2B is $3(0.51)^2(0.49)$.

Without going into further details we can now see the simple mechanism by which one obtains the probabilities for the various family make-up in a series of any number of births. For example, let us consider we have four births and we wish to find the probability of the event of two boys and two girls. This event is made up of 6 cases, namely

$$GGBB$$
$$GBGB$$
$$GBBG$$
$$BGGB$$
$$BGBG$$
$$BBGG$$

The probability of each of these cases is $(0.51)^2(0.49)^2$. Since these six cases make up the event, the probability of the event $2B$ is $6(0.51)^2(0.49)^2$. Here we found the number 6 by actually listing the cases which had two boys and two girls in any order. However there are easier ways to find the number of cases in any event.

Pascal's triangle

One such way is by use of *Pascal's Triangle*. In the first row one writes the number 1, and in the second row the numbers 1,1 so that the three numbers form a triangle. Then one constructs additional rows of this triangle by adding together two adjacent numbers in a row and placing the sum in the next row down at a location between the two adjacent numbers. In the application of this rule it is assumed that rows are extended by zeroes to the left and to the right. Thus Pascal's Triangle for the first six rows is

```
Row 0                    1
Row 1                  1   1
Row 2                1   2   1
Row 3              1   3   3   1
Row 4            1   4   6   4   1
Row 5          1   5  10  10   5   1
```

The top row is labeled 0, the next is row 1, the next row 2, etc. We see that row three, namely, 1, 3, 3, 1, gives for three births the number of cases for the events $0B$, $1B$, $2B$, and $3B$ respectively. Likewise row number four, namely, 1, 4, 6, 4, 1, gives for four births the number of cases for the events $0B$, $1B$, $2B$, $3B$, and $4B$ respectively.

This table can be easily continued. Row number 6 is obtained by $0 + 1 = 1$, $1 + 5 = 6$, $5 + 10 = 15$, $10 + 10 = 20$, $10 + 5 = 15$, $5 + 1 = 6$, $1 + 0 = 1$. Because of these calculations, we see that the sum of the numbers in row 6 is twice the sum of numbers in row 5. In fact, we have

$$\text{Sum of Numbers in Row } 0 \ = \ 2^0 = 1$$
$$\text{Sum of Numbers in Row } 1 \ = \ 2^1 = 2$$
$$\text{Sum of Numbers in Row } 2 \ = \ 2^2 = 4$$
$$\text{Sum of Numbers in Row } 3 \ = \ 2^3 = 8$$
$$\text{Sum of Numbers in Row } 4 \ = \ 2^4 = 16$$

$$\cdots$$

$$\text{Sum of Numbers in Row } n \ = \ 2^n = \underbrace{2 \cdot 2 \cdot 2 \cdots 2}_{n \text{ times}}$$

The various rows of the table are symmetrical. In rows corresponding to an even number of births, say row 4, the numbers increase toward the middle, and the greatest number occurs at the middle corresponding to an equal number of boys as girls. In rows corresponding to an odd number of births, the numbers also increase toward the middle, but now there are two equal numbers in the middle corresponding to the two cases where there is one more boy than girl or one more girl than boy.

Each number in Pascal's Triangle can be represented by a symbol. This symbol is made up of two numbers one over the other and both enclosed by a set of parentheses. The top number is the number of births and the bottom number is the number of boys. With this symbol Pascal's Triangle becomes

$$\binom{0}{0}$$

$$\binom{1}{0} \quad \binom{1}{1}$$

$$\binom{2}{0} \quad \binom{2}{1} \quad \binom{2}{2}$$

$$\binom{3}{0} \quad \binom{3}{1} \quad \binom{3}{2} \quad \binom{3}{3}$$

which corresponds to

$$
\begin{array}{ccccccc}
 & & & 1 & & & \\
 & & 1 & & 1 & & \\
 & 1 & & 2 & & 1 & \\
1 & & 3 & & 3 & & 1 \\
\end{array}
$$

In general the symbol for the number of cases of n births resulting in x boys is

$$\binom{n}{x}$$

Permutations and combinations

The mathematical term for the symbol $\binom{n}{x}$ is the number of combinations of n things taken x at a time. In order to understand combinations we must first discuss permutations.

Each different arrangement either of all or part of a number of things is called a permutation. Thus there are two permutations of the two letters A and B, namely AB and BA. There are six permutations of the three letters A, B, and C, namely ABC, ACB, BAC, BCA, CAB, CBA. There are also six permutations of the three letters A, B, and C taken two at a time, namely AB, AC, BA, BC, CA, CB.

Let us now find an expression for the number of permutations of n things taken x at a time, symbolized by P_x^n. The first place can be filled in any one of n ways. This leaves $n - 1$ ways of filling the second place by any one of the $n - 1$ remaining things. The third place can be filled in $n - 2$ ways, the fourth in $n - 3$ ways, and so on until finally the xth place in $n - x + 1$ ways. Hence the number of ways of filling the x places is

$$P_x^n = n(n - 1)(n - 2)(n - 3) \cdots (n - x + 1)$$

We observe that this expression can also be written as

$$P_x^n = \frac{n!}{(n - x)!} = \frac{n(n - 1)(n - 2)(n - 3) \cdots (3)(2)(1)}{(n - x)(n - x - 1)(n - x - 2) \cdots (3)(2)(1)}$$

If $x = n$ then the expression for the number of permutations is

$$P_n^n = n(n - 1)(n - 2)(n - 3) \cdots (3)(2)(1) = n!$$

That is, the number of permutations of n things taken n at a time is $n!$ ($= n$ factorial).

Some slight knowledge of permutations seems to have existed among the ancient Chinese. Although the Greeks gave the subject little attention, the Latin mathematician Boethius in about 510 AD gave the rule for the number of permutations of n things taken two at a time, that is

$$P_2^n = n(n - 1)$$

Finally, about 1150 AD, Bhāskara gave the general rule for both permutations and combinations.

A combination is a group of things that is independent of the order of the various distinct elements in the group. For example, the musical effect of striking three notes at the same time on a piano, say C, E, G, is the same whether we think of the notes as in one order or another. This is a combination of notes. If, however, we strike the same notes in succession, the musical effect is different when the notes are struck in different orders. This is a permutation of notes.

Let us now find an expression for $\binom{n}{x}$, the number of combinations of n things taken x at a time. Since any change of order of the elements in a combination does not alter the combinations, it follows that in any one combination of x things there are $x!$ permutations. Hence there are $x!$ times as many permutations in a given group as there are combinations. That is,

$$P_x^n = x! \binom{n}{x}$$

or

$$\binom{n}{x} = \frac{P_x^n}{x!} = \frac{n(n - 1)(n - 2) \cdots (n - x + 1)}{x(x - 1)(x - 2) \cdots (1)}$$

In the application of this formula we observe that there are the same number of factors in the numerator as there are in the denominator.

An alternative expression is

$$\binom{n}{x} = \frac{n!}{x!(n-x)!}$$

As an example, the number of cases of 4 births resulting in 3 boys is the number of combinations of 4 things taken 3 at a time, namely

$$\binom{4}{3} = \frac{4!}{3!1!} = \frac{4 \cdot 3 \cdot 2 \cdot 1}{3 \cdot 2 \cdot 1 \cdot 1} = 4,$$

which is the same result that we can obtain from Pascal's Triangle.

The binomial distribution

We can summarize our results up to this point by saying that in n births the number x of boys follows a binomial distribution. As we have seen the probability of each case with x boys and $n - x$ girls is

$$(0.51)^x(0.49)^{n-x}$$

and there are $\binom{n}{x}$ such cases. Hence in n births the probability of x boys is

$$P(x) = \binom{n}{x} (0.51)^x(0.49)^{n-x}$$

This equation represents the *binomial distribution*, where the number x of boys can take any value $x = 0, 1, 2, \ldots, n$. We have already written down the distributions when $n = 1, 2,$ and 3, which were

$n = 1$	$P(0)$	$=$	0.49		
	$P(1)$	$=$	0.51		
			Total 1.00		
$n = 2$	$P(0)$	$=$	$(0.49)^2$	$=$	0.24
	$P(1)$	$=$	$2(0.51)(0.49)$	$=$	0.50
	$P(2)$	$=$	$(0.51)^2$	$=$	0.26
				Total	1.00
$n = 3$	$P(0)$	$=$	$(0.49)^3$	$=$	0.12
	$P(1)$	$=$	$3(0.51)(0.49)^2$	$=$	0.37
	$P(2)$	$=$	$3(0.51)^2(0.49)$	$=$	0.38
	$P(3)$	$=$	$(0.51)^3$	$=$	0.13
				Total	1.00

Let us now change our terminology from that of a family to the general terminology for the binomial distribution.

Instead of a birth we speak of a *trial* of an experiment. Each trial has two possible outcomes, namely *success* or *failure*. The probability of a success on a given trial is denoted by p and the probability of a failure on the given trial is denoted by q. Since a trial must result in either a success or a failure we have $p + q = 1$. In our case success corresponds to the birth of a boy B and the probability of a success is $p = 0.51$. Failure (and the use of the word is only as an expression) corresponds to the birth of a girl G and the probability of failure is 0.49. The binomial distribution gives for n trials the probability of x successes. This probability is

$$P(x) = \binom{n}{x} p^x q^{n-x}$$

where x can vary over the range 0, 1, 2, . . . , n. It is customary to say that the number of successes in n trials is a random variable x having the binomial distribution. This distribution applies whenever the probability of a success remains constant from trial to trial and the trials are independent.

Mean of the binomial distribution

Let us now apply the concept of expectation to the binomial distribution. First consider the distribution of the number x of boys in the case of one birth:

$$x = 0 \qquad P(x = 0) = 0.49$$
$$x = 1 \qquad P(x = 1) = 0.51$$

We multiply the number of boys by the corresponding probability to find the expectation of each event. By adding all these products we obtain the expected number of boys:

$$0 \cdot P(0) + 1 \cdot P(1) = 0.51$$

Thus in one birth the expected number of boys is 0.51. We call this expected number the mean of the binomial distribution for $n = 1$. Note that this mean is equal to $np = (1)(0.51)$.

Next consider the binomial distribution for $n = 2$ births:

Number of boys	Probability	Product
$x = 0$	$P(x = 1) = 0.24$	0.00
$x = 1$	$P(x = 2) = 0.50$	0.50
$x = 2$	$P(x = 3) = 0.26$	0.52

We take each product of the number of boys times its probability. We sum these products to find the expectation:

$$0.50 + 0.52 = 1.02$$

Thus the mean number of boys in 2 births is 1.02. Note that this mean is equal to $np = 2(0.51) = 1.02$.

Now consider the binomial distribution for $n = 3$ births:

Number of boys	Probability	Product
0	0.12	0.00
1	0.37	0.37
2	0.38	0.76
3	0.13	0.39

| | Total | 1.52 |

The expected number of boys in 3 births is the sum of the products, namely 1.52. Because we rounded off our probabilities to two decimal places, there is round-off error in our expected value. If we had carried more decimal places, the expected value would have been 1.53, which is $np = 3(0.51)$.

We see that the *mean of a binomial distribution* is equal to np. In other words the mean of a binomial distribution is equal to the product of the number of trials and the probability of success on an individual trial. Another term for the mean is the expected value of the number of successes. Hence we may write that for a binomial distribution

$$Ex = np$$

which is an abbreviation for "the expected value of x is equal to np." Often we denote the mean of a probability distribution by the Greek letter μ. Hence we may also write that for a binomial distribution

$$\mu = np.$$

We have not proved this formula that the mean of a binomial distribution is given by np, but the formula is intuitively reasonable. For example, in 100 births we would expect $np = 100(0.51) = 51$ boys.

In this section we have seen how to find the mean or expected number of boys in n births. This mean is a measure of the central location, and is the counterpart of the mean or average of a set of data. In fact the formula for the *empirical mean* $\bar{x}$ is

$$\bar{x} = \Sigma xf$$

where x represents the variable and f represents its relative frequency. The formula for the theoretical mean μ is

$$\mu = \Sigma xP$$

where x represents the random variable and P represents its probability. Even as the relative frequency f is an estimate of the probability P, so is the empirical mean $\bar{x}$ an estimate of the theoretical mean μ. In the case of the binomial distribution we have exact expressions for the probability values P and hence we can theoretically compute the value of μ which turns out to be np.

Variance of the binomial distribution

Previously we have seen how to describe the variability of data expressed in terms of an empirical frequency distribution. In particular the *empirical variance* is given by the formula

$$s^2 = \frac{1}{n-1} \Sigma(x - \bar{x})^2 f$$

where f is now the absolute frequency (i.e. the numbers of data points in each class interval). In case the theoretical mean μ is known, we should instead use the formula

$$s^2 = \frac{1}{n} \Sigma(x - \mu)^2 f$$

That is, when the empirical mean $\bar{x}$ is used, the divisor in s^2 is $n - 1$, whereas if the theoretical mean μ is used the divisor is n. Since relative frequency is absolute frequency divided by n, this last formula in terms of relative frequency is simply

$$s^2 = \Sigma(x - \mu)^2 f$$

The formula for the theoretical variance is

$$\sigma^2 = \Sigma(x - \mu)^2 P$$

where P is the counterpart of f and σ^2 is the counterpart of s^2. The quantity σ^2 is called the *(theoretical) variance*. It is represented as the square of the Greek lower case letter sigma σ. The theoretical variance may also be described as the expected value of $(x - \mu)^2$. That is,

$$\sigma^2 = E(x - \mu)^2$$

Let us now compute the variance of the binomial distribution for one birth. We have:

$$n = 1 \qquad \mu = 0.51$$

Number of boys	Probability	Squared Deviations	Products
0	0.49	$(0 - 0.51)^2$	0.128
1	0.51	$(1 - 0.51)^2$	0.122
		Total	0.250

The total of the products of squared deviations times probability is

$$\sigma^2 = 0.250$$

We see that this value is $npq = (1)(0.51)(0.49) = 0.250$.

Let us next compute the variance of the binomial distribution of two births. We have (where $n = 2$, $\mu = 1.02$)

x	P	$(x - \mu)^2$	$(x - \mu)^2 P$
0	0.24	$(0 - 1.02)^2$	0.25
1	0.50	$(1 - 1.02)^2$	0.00
2	0.26	$(2 - 1.02)^2$	0.25
		Total	0.50

The total is the variance $\sigma^2 = 0.50$ which again we see is equal to $npq = 2(0.51)(0.49)$.

Finally, the binomial distribution for $n = 3$ births is (where $\mu = 1.53$):

x	P	$(x - \mu)^2$	$(x - \mu)^2 P$
0	0.12	$(0 - 1.53)^2$	0.281
1	0.37	$(1 - 1.53)^2$	0.104
2	0.38	$(2 - 1.53)^2$	0.084
3	0.13	$(3 - 1.53)^2$	0.281
		Total	0.750

The total is the variance $\sigma^2 = 0.750$, which again we see is $npq = 3(0.51)(0.49)$.

In general we can say that the *mean μ* of a *binomial distribution* is np and the *variance σ^2* is npq. As a result the *standard deviation* of a binomial distribution is the positive square root of the variance:

$$\sigma = \sqrt{npq} .$$

The standard deviation of the above three distributions are:

$$\text{For } n = 1, \quad \sigma = \sqrt{0.25} = 0.50$$
$$\text{For } n = 2, \quad \sigma = \sqrt{0.50} = 0.71$$
$$\text{For } n = 3, \quad \sigma = \sqrt{0.75} = 0.87.$$

Independence between individual births

During the last century in some parts of Europe a man whose wife was expecting a baby would be extremely anxious for a boy. He would go to the registrar of births in the town to see how many boys and how many girls were born in the last few days. If the number of girls was comparatively large, he would be happy and would feel that his probability of having a son instead of a daughter was exceedingly good.

Because of the great interest in this problem, statistical evidence was collected. Statistical data from four towns in Bavaria showed that out of 200,000 births there was only one run of 17 consecutive births of babies of the same sex, and not a single such run of greater length.

Many people interpret this statistical data as follows. Once 17 girls are born in succession in a town, the probability that the next baby born is a boy is so great that the event is practically a certainty. Thus they reason: After a run of 17 consecutive girl births, the next birth is for all intents and purposes a sure thing, namely the certain birth of a boy, instead of a random event.

But what about the man who was the father of the 17th girl in this run of 17 consecutive girls? Before his daughter was born, he also went to the town registrar and saw that there was a run of 16 consecutive female births. Was he not happy also, and did he not consider that his probability of a male birth was exceedingly good? And yet fate decreed otherwise, and his wife had a girl, despite the fact that the 16 preceding births were also girls.

As more and more statistical evidence is collected and assimulated the following characteristic stands out. Let us denote the birth of a boy by B and birth of a girl by G. Then the successive births would be represented by a series, such as

$$\ldots BBGGGGBGBGGBGBGBBBG \ldots$$

The statistical evidence shows that the distribution of B's and G's in the partial series formed by all births following 17 consecutive G's is the same as the distribution in the whole series. Consequently on the basis of this evidence, the probability of a boy after a run of 17 consecutive girls is the same as the probability of a boy regardless of what births came before. In other words, the probability of a boy is not altered by the event of the 17 preceding births all being girls.

Let

$$G^{17}$$

designate the event of 17 consecutive girls. Let

$$B|G^{17}$$

designate the event of a boy given that the 17 preceding births were all girls. (Note: The vertical bar | is read *given that*, or simply *given*.) Let

$$B$$

designate the event of a boy.

Now let X designate any arbitrary event. Then we let

$$P(X)$$

designate the *probability* that the event X happens.

Thus

$$P(G^{17})$$

is the *probability* that the event "17 consecutive girls" happens, and

$$P(B|G^{17})$$

is the *probability* that the event "a boy given that the 17 preceding births were girls" happens. And finally

$$P(B)$$

is the *probability* that the event "a boy" happens.

Then the empirical experience of mankind, based on statistical evidence collected over the centuries, indicates the following relationship holds

$$P(B) = P(B|G^{17}).$$

That is, the probability of a boy is the same as the probability of a boy given that the 17 consecutive prior births were girls. Thus we may say that the event "a boy" is independent of the event of 17 consecutive prior girl births.

Exercises

1. Seven horses are in a race. In how many ways can three from among them finish first, second, and third? *Answer.* 210

2. How many five-card hands can be dealt from a 52-card deck? *Answer.* 2,598,960

3. How many three-letter words can be made from ten different letters (a) if repeats are allowed? (b) if repeats are not allowed? (c) if a letter may be repeated only once? *Answer.* (a) 1000 (b) 720 (c) 990

4. (a) In how many ways can a president, a secretary, and a treasurer be selected from a club of 20 people? (b) In how many ways can a committee of three be selected from the club? *Answer.* (a) 6840 (b) 1140

5. In how many ways can the 26 letters of the alphabet be lined up so that A and B are adjacent? *Answer.* $2 \cdot 25!$

6. (a) How many 13-card bridge hands can be dealt from a deck of 52 cards? (b) In how many ways can 13-card hands be dealt to North, South, East, and West?

$$\text{\textit{Answer}} \quad \text{(a)} \ \binom{52}{13} \quad \text{(b)} \ \binom{52}{13}\binom{39}{13}\binom{26}{13}\binom{13}{13}$$

7. How many five-card poker hands consist of (a) two pairs? (b) a full house? (c) a straight flush? (d) four of a kind?

$$\text{\textit{Answer}} \quad \text{(a)} \ \binom{13}{2}\binom{4}{2}\binom{4}{2} \cdot 44 = 123{,}552$$

$$\text{(b)} \ 13 \binom{4}{3} 12 \binom{4}{2} = 3774$$

(c) $4 \cdot 10 = 40$
(d) $13 \cdot 48 = 624$

8. A lock has 40 positions. A "combination" for the lock consists of four settings, and no setting can coincide with the preceding one. How many such "combinations" are there? *Answer.* $(40)(39)(39)(39) = 2,372,760$

9. Six people sit in six chairs in a circle. If everyone moves the same number of places to the left, the seating is considered the same as before. How many different seatings are there? *Answer.* $6!/6 = 120$.

10. An entering college student must take one of five science courses, one of six history courses, one of five English courses, and one of three mathematics courses. How many programs are available to him? *Answer.* $(5)(6)(5)(3) = 450$

11. An advertising agency claims that, of the college students who smoke, 25 percent smoke brand A. If we take a random sample of four students who smoke, and if the claim is true, what is the probability that at least one of them will be found to smoke brand A? *Answer.* 175/256.

12. In Boston, rain falls on the average one day out of every three days during which the sky is overcast. What is the distribution of the number of days with rainfall among the next three overcast days, assuming complete independence? Find its mean and variance.

$$\textit{Answer.} \quad P(x) = \binom{3}{x}\left(\frac{1}{3}\right)^{x}\left(\frac{2}{3}\right)^{n-x} \text{ for } x = 0, 1, 2, 3.$$

$$\text{Mean} = 1, \text{ Variance} = 2/3$$

13. In a large orchard, 10 percent of the apples have worms. If four apples are picked at random, what is the probability that (a) exactly one will be wormy? (b) none will be wormy? (c) at least one will be wormy?

Answer. (a) 0.2916 (b) 0.6561 (c) 0.3439

CHAPTER 6

POPULATION AND SAMPLE

Population

Any set of individuals, objects, or ideas that have some common observable characteristic makes up a *population* or *universe*. For example, the set of all people living in America would be a population of individuals; the set of all electric light bulbs in a warehouse would be a population of objects; the set of all poker hands (i.e. a set of 5 cards dealt from a shuffled deck) is a population of ideas. The notion common to all populations is that of aggregation; the term population is employed to denote any collection of a specific type under consideration.

A population can be finite or infinite. The population of people in America is finite. The population of electric light bulbs in a warehouse is finite. Moreover the population of people in America is existent; that is, the population exists in the sense that we can go out and count each member of the entire population. Likewise the population of electric light bulbs is existent; only lack of time, money or opportunity would prevent us from examining the whole population. On the other hand, the population of all poker hands is not finite. Each time we gather up the cards, reshuffle the deck, and deal a new hand we generate another member of the population. Now it is clear that no matter how many times we deal we can never obtain the complete population, for we can always add new members by another deal. As a result this population is infinite. Furthermore, the population does not exist in the sense that we can go out and look at any member. Instead the members have a kind of hypothetical existence conferred on them by the notion of the process of dealing a hand.

With the above comments in mind, we may say that generally a population is an abstract set, which in certain specific cases actually reduces to a concrete set. Let us give some further examples: the population of a shipload of grain and the population of points on a rifle target. The shipload of grain is finite but so large that we usually would treat it as infinite; the points on a target are indeed infinite. Most of our procedures will be based on the assumption of an infinite population.

One *definition* of the science of *statistics* is that it is the branch of scientific method which deals with the properties of populations. More specifically statistics deal with the numerical properties of populations, so with each member of the population we associate a number called the measurement. Depending on the characteristic at hand, the numbers can be continuous or discrete. For example, if we consider the height of people in America the variable "height" is a continuous variable; it can take on all possible values from say one-half to eight feet. In dealing with continuous variables we never look at one particular value, say five feet eight inches, but at a small interval around a particular value, say five feet eight inches give or take one-half inch. In other words there is always some region of tolerance associated with any measurement of a continuous variable. As another example, if we consider the number of children that each women has in America, the variable "number of children" is a discrete variable; it can take on only the discrete values 0, 1, 2, . . . , 20 or whatever the upper limit might be.

Sample

Any subset of a population is a *sample* from that population. The *size of a sample,* usually represented by the letter *n,* is the number of members in the sample.

A field of corn has an area of 1000 acres. A sample of 20 plots of one acre each is chosen, and the yields of these plots are measured. The population consists of the 1000 individual acres as members. The sample is the 20 members chosen, while the measurement is the yield of the plot.

An investigation is made to test the effects of a certain type of vaccine for flu. A group of subjects infected with flu are treated with the vaccine, and the number recovering from the flu within a specified time interval is observed. The sample consists of the group of subjects actually used, and the measurement would be the label either "recovered" or "did not recover." The population here could be considered to be all people with the flu.

A study is made as to the age of people in the United States. A sample of 1000 people is chosen, and their ages recorded. Here the population consists of all people in the United States. The sample consists of the 1000 people chosen. The measurement taken is age.

Random sampling

A critical factor in finding out what use can be made of a sample is the method used in choosing the sample. If some members of the population have a greater chance to be chosen than others, then we say the sample is biased. For example, suppose we are sampling people to determine the average age of a population. If each time we came upon an older looking person we neglected him and instead picked a younger looking person, then our sample would be biased toward younger people. Such subjective methods of picking members from a population for a sample often result from subconscious or sometimes conscious preferences of the person making the selections. In order to prevent such bias, an objective method of picking a sample ought to be employed.

When every member of the population has an *equal* and *independent* chance of being chosen for a sample, the sample is called a *random sample.* In other words, a random sample is one in which (1) all members of the population have an equal opportunity to be drawn into the sample, and (2) each member is selected independently of whether any other member is drawn into the sample. Taken together, these conditions mean that any group of members is as likely to be chosen as any other group of the same size.

Technically speaking, every member chosen should be measured and returned to the population before another selection is made. As a result it is possible that some member can be chosen twice in the same sample. For example, to choose a random sample of 5 cards out of a population of 52 cards, you ought to choose one card, record its value, return the card to the deck, shuffle the deck, then draw the second card, record its value, return it to the deck, shuffle, and so on until 5 cards have been recorded. If the population is large compared to the sample size, then only a small error will result from the procedure of not returning each sampled member back to the population. In practice it is usual not to return each sampled member, and in fact sometimes it is impossible to do so, as in a case where a member is changed or destroyed in the sampling process.

Theoretical frequency curve (for a finite population)

Note that we said in our definition of random sample that every member of the population has an equal chance of being in the sample. We did not say, and it is wrong to say, that every measurement in the population has an equal chance of being in the sample. For example, in the U.S. population there are many more people whose height measures 5'9" than 6'9". In a random sample each member has an equal opportunity, and as a result the measurement 5'9" will have a greater chance of being drawn than 6'9". In fact the curve that represents the probabilities of the measurements for a single draw (i.e. a sample of size one) is the *theoretical frequency curve* for heights in the population.

The theoretical frequency curve is an abstraction from the empirical frequency curve of a set of data. If the set of data makes up the measurements of the entire population (as it can only in the case of certain finite population) then the theoretical and empirical curves are the same. For example, the U.S. Census each ten years records the age of each person in the United States. The empirical frequency curve of ages for this data, because it exhausts the entire U.S. population, is the theoretical distribution as well. The median of this distribution is that age for which half the people are older and the other half younger. Now we are going to make an important jump ahead in our statistical logic. We are going to say that if we draw a person at random (that is, if we take a random sample of size one) then there is a fifty percent chance that this person will be older than the median age and a fifty percent chance this person will be younger than the median age. In other words, our statistical logic is this: Fifty percent of the people in the population are older than the median age; therefore, a person drawn at random has a fifty percent chance of being older than the median age. The above reasoning is general; the median is associated with 50 percent, but one can use any other benchmark with its associated percent. For example, the concept can be extended from one value (the median) which divides the total frequency into two equal parts, to an arbitrary value which divides the total frequency into two unequal parts. The percentage that lies above this value becomes the chance that a random draw exceeds this value. Likewise the percentage that lies below this value becomes the chance that a random draw falls short of this value.

Theoretical frequency curve (for an infinite population)

Let us consider a simple example. Suppose that a coin is tossed a number of times and frequencies of heads and tails are recorded. The relative frequencies of each can be shown in a histogram. In the figure we show three histograms, one for 10 throws, one for 100 throws, and one for 1000 throws. Note that on the abscissa we represent tails by 0 and heads by 1.

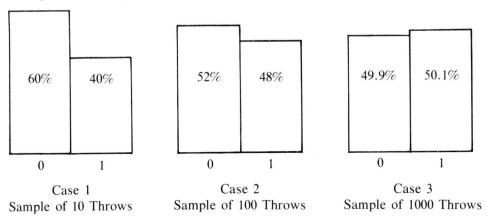

Case 1	Case 2	Case 3
Sample of 10 Throws	Sample of 100 Throws	Sample of 1000 Throws

As the number of throws is increased so in the regularity of the histogram, and the frequency distribution of the sample is said to tend to the frequency distribution of the population. In this case, for an honest coin, the frequency distribution of the population is 50% − 50%. Note here that the *a priori* (i.e. beforehand) knowledge of the population distribution allows us to predict with reasonable accuracy the behavior of the sample.

As a further demonstration, let us take the total score when 10 coins are thrown at the same time. This score cannot be less than zero or more than 10. Again, the frequency distribution of the population can be determined by using the methods of probability theory. The figure shows how the histograms become more regular as the throws increase, and tend to the frequency distribution of the population:

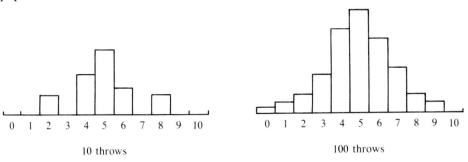

10 throws   100 throws

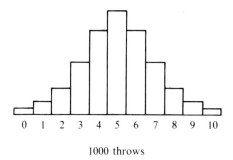

1000 throws

The essential feature of a histogram is that the proportion of occasions on which the score is between any two values is represented by the area of the blocks of those two values and all the blocks in between.

A more practical example is shown in the figure which gives histograms for heights of American men:

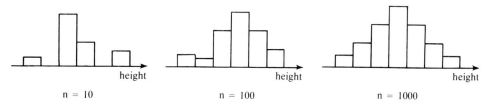

As the size of the sample is increased the irregularities of the histogram disappear and the form of the frequency curve of the population becomes more apparent. Of course the distribution is not determined *a priori* (beforehand) as with the distribution of scores of coin tossing. However this distribution of heights and the distribution from coin tossing bear marked similarities. This similarity is not purely coincidental but is a consequence of the *central limit theorem* which determine the shape of many frequency distributions experienced in practice, namely the *normal curve*. More on the central limit theorem and the normal curve will be given in the next chapter.

In the examples given on the tossing of coins, the scores must be whole numbers. Such cases where the measurement must be a whole number do occur, as for example, when counts of insects, plants, animals, or inventory are made, but it is more usual for measurements to be able to take on all values between two limits and not be confined only to whole numbers. For example, all heights between four feet and eight feet are possible because men's heights do not increase by jumps of exactly one inch with nothing in between. Measurements such as this are said to be continuous.

Grouping was introduced in the first chapter as a method of comparing the relative frequencies of various measurements. In order to make the comparison worthwhile it was necessary to choose the grouping interval sufficiently large to include several measurements. As we increase the size of the sample, however, the grouping interval can be made smaller. In fact for a large enough sample, the grouping interval can be made so small that the blocks of the histogram blend together and appear as a curve. That is, the theoretical frequency curve results from the process of indefinitely increasing the same size and thereby reducing the grouping interval indefinitely.

The important point that we want to make is the following. *The theoretical frequency curve retains the property of the histogram in that the area under the curve between any two values gives the frequency between the two values.* Usually the histogram is plotted in terms of *relative frequencies*. Then the theoretical frequency curve is such that the *area* between any two values gives the *probability* that a random draw will occur between those two values. Two terms often used are *random variable* and the *probability density* of the random variable. The numerical value of the random draw represents the random variable, and the theoretical (relative) frequency curve represents its probability density.

Exercises

1. In each of the following examples, say whether the population is existent or conceptual: all ball-bearings that can be turned out by a given machine; all the whales in the sea; all wheat plants; all bridge hands in card playing; all the coal reserves of the world; all people that like music; the weather in Boston; the flight of a wild goose; next year's orange crop.

2. In each of the following examples, say whether the population is finite or infinite: all poker hands in card playing; all hydrogen atoms; all the stars; all people; automobiles.

3. Toss a die 10 times and draw the histogram; toss the die 90 more times and draw the revised histogram.

4. Discuss methods of obtaining random samples (with replacement) from the following populations, and give for each a possible measurement that might be made on the members in the sample: (a) Apples on a tree, (b) People in a state, (c) Students in a university, (d) A month's production of automobile tires, (e) Fish in a lake, (f) Seeds planted in a field.

5. Suppose that we wish to determine the average number of people in a family in Boston. Would the recording of the number of people in each student's family in a random selection of students at the University of Massachusetts in Boston be a reasonable way to obtain the necessary data?

6. What possible bias could result from a sample obtained by selecting every tenth item in a population? Why is this not a random sample?

7. Suppose we ask whether two school classes are significantly different, based on the results of some tests. In what populations are we interested?

8. Suppose there are 10,000 students in attendance at a given university, and a test has been given to 400 of these students. The complete set of 10,000 students we could have tested is called a

(a) mean	(c) population
(b) sample	(d) statistic

9. The statement "Mary's dancing is above average" means Mary's dancing

(a) represents some middle position or value
(b) is mutually exclusive
(c) on some scale is above some middle position or value
(d) is the mode

10. Let k represent any one of the set of numbers 1, 2, 3, 4, 5, 6, 7, 8, 9. The subset $3 < k < 6$ is
(a) 3, 4, 5, 6   (c) 4, 5, 6
(b) 3, 4, 5   (d) 4, 5

11. Which of the following variables would you regard as continuous?
(a) number of students in each class at a university
(b) number of children per married couple
(c) time spent in preparing an assignment
(d) number of correct responses on a test

12. Which of the following variables would you regard as discrete?
(a) reaction time to stop a car
(b) annual income of professors at a university
(c) speed of running a maze
(d) time spent waiting for an elevator

13. Which word does not belong with the others below?
(a) data point   (c) sample value
(b) statistic   (d) parameter

14. When tossing a perfect die each face is
(a) equally likely   (c) of zero probability
(b) biased   (d) of probability one

15. The population mean μ can be found by means of a sample provided the sample size
(a) is greater than 1000   (c) is finite
(b) exhausts the whole population   (d) in the 95% range

16. From sample to sample, the theoretical frequency distribution is
(a) dependent upon sample size
(b) the same
(c) the same as the empirical frequency distributions
(d) the most variable of the empirical frequency distributions

17. For a random sample with replacements, each sample member is
(a) except for the first, dependent on the previous sample member
(b) dependent upon all of the other sample members
(c) equally likely and independent
(d) equal to every other sample member

18. In order to keep a sample random, an extreme data point ought to be
(a) thrown out   (c) reduced
(b) retained   (d) emphasized

CHAPTER 7

THE NORMAL DISTRIBUTION

Background

One of the most important theoretical frequency distributions in statistics is the *normal distribution*. Many statistical procedures are based on a knowledge of this distribution. Unlike most other important geometrical forms, the shape of the normal curve does not appear in any common object familiar to the eye. The closest object that resembles a normal curve is the outline of a bell. The normal curve is often described as a symmetrical bell-shaped curve, extending infinitely far in both positive and negative directions.

The concept of the normal distribution is one that has universal appeal. Nearly all scientists believe in the normal distribution: Observers because they believe it is a theorem of mathematics; mathematicians because they believe that it is aesthetically satisfying; aestheticians because they believe it is philosophically true; philosophers because they believe it is a fact of observation. A few neither believe it nor disbelieve it; these maintain that the normal distribution is a convenient statistical procedure for the bookkeeping of data either in raw form or under suitable transformations, and about bookkeeping procedures it ought to be asked, not if they are true or false, but are they useful.

The central limit theorem and the normal curve

It can be shown mathematically that whenever a measurement is the sum of a large number of small independent effects, no one of which predominates, the distribution of the measurement will take approximately the same general bell-like shape. This bell-like shape is the normal curve, and a knowledge of its central value (mean) and spread (standard deviation) determines the curve completely. That is, once the mean and standard deviation are known all else is determined. The mean and standard deviation are called *parameters,* so the normal curve is determined by two parameters. Since the normal distribution is a theoretical frequency curve, it is customary to use Greek letters to represent its parameters. More specifically, the Greek letter μ (called mu) denotes its mean, and the Greek letter σ (called sigma) denotes its standard deviation. The figure illustrates the normal curve with mean μ and standard deviation σ.

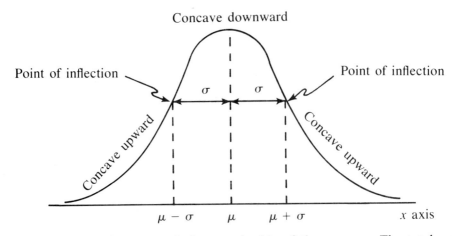

The normal curve is symmetrical on each side of the mean μ. The total area between this curve and the x-axis is one square unit. The curve is concave downward for x within $-\sigma$ to σ of the mean, and concave upward for x outside of this range. As a result there are *two points of inflection,* one at $\mu - \sigma$ and the other at $\mu + \sigma$. (A point of inflection is the point where the slope stops getting steeper and starts getting flatter as you slide down a hill.) Thus the standard deviation is the physical horizontal distance from the mean to either point of inflection. (Just as all circles have the same general shape and are determined by the center and radius, all normal curves have the same general shape and are determined by the mean and standard deviation. The center of the circle and the mean of the normal curve are location parameters, whereas the radius of the circle and the standard deviation of the normal curve are spread parameters. As in its class the circle is conceptually the simplest possible geometric shape, so also in its class the normal curve is the simplest possible geometric shape. One might say that the normal curve plays a role in statistics that is analogous to the role played by the circle in ancient mathematics and astronomy.)

As we have just stated, the normal distribution results (in the limit) in the case of any variable made up as a sum of many small independent variables as long as each small independent variable has a negligible effect on the sum. That is, no one of the variables in the sum is allowed to have a predominant effect on the sum. This remarkable result is called the *central limit theorem.*

Let us now look at some examples of the central limit theorem. The total consumption of electric power delivered by the Edison Company is the sum of the quantities consumed by the various customers, so the total consumption has a normal distribution. Likewise we would expect a normal distribution for the total gain or loss on the risk business of an insurance company, as the total is the sum of the gains or losses on each single policy. We would expect the total error committed in a physical or astronomical measurement to have a normal distribution, as the total error is the sum of a large number of mutually independent elementary errors. The yield of wheat on a farm would be normally distributed, as it is the result of many small effects due to environmental factors such as rainfall, temperature, soil fertility, pest activity, and plant competition and to a multitude of genetical factors. The distribution of the result of 10 tosses of a coin is approximately normal, because the total score is the sum of ten independent scores resulting from each toss.

Despite this seeming universality of the normal distribution, we cannot expect it to apply to every measurement for various reasons. First, there may not be a large number of small effects. Second, one particular effect may predominate. Third, the effects may not be independent, as with rainfall and temperature, although a small degree of dependence may not seriously effect the normality provided there are a large enough number of other independent effects. Fourth, the effects may not be additive.

The standard normal distribution

There are an infinite number of normal distributions. However all normal distributions have the same general shape, namely the shape of a bell, and they differ from each other only with respect to their means and standard deviations. A particular normal distribution is completely determined by the specification of its two parameters: its mean μ and its standard deviation σ. We say that the normal distribution represents a two parameter family of curves and both parameters must be assigned to specify a particular normal curve. If the standard deviation σ is fixed and if we vary the mean μ, we have a family of curves with identical shape but with different locations on the horizontal axis. If the mean μ is fixed and if we vary the standard deviation σ, we have a family of curves with identical location on the horizontal axis but with different spreads.

One of the main uses of the normal curve is to calculate probabilities. For this purpose we direct our attention on one particular normal distribution whose mean is 0 and whose standard deviation is one. This distribution is called the *standard normal distribution*. The abscissa for the standard normal distribution is labeled z, and z is called the *standard normal variable*. Probabilities are assigned to intervals of possible values of the variable z, rather than to single values of z. *The probability that on a single draw the variable z will fall within a stated interval is equal to the area under the curve within that interval.* For example, the area under the curve within the interval $-1.96 < z < 1.96$ is 0.95. Thus the probability that a value of z drawn at random from the population will fall in the interval $-1.96 < z < 1.96$ is 0.95. We illustrate this result by the figure

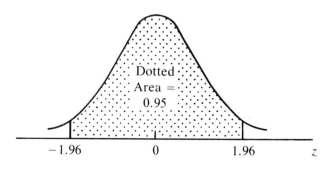

or by the equation

$$P(-1.96 < z < 1.96) = 0.95$$

which we read as "The probability of z between -1.96 and 1.96 is 0.95."

The probability that an observed z lies between $-\infty$ and ∞ is one. We can write this statement as

$$P(-\infty < z < \infty) = 1.00$$

which states that the area under the standard normal curve is equal to one.

The essential thing to remember is: The probability of z falling into any given interval is equal to the area under the curve for that interval.

The standard normal distribution is the curve shown in the figure

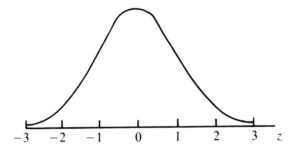

We see that the standard normal curve is symmetric about $z = 0$. That is, the portion of the curve to the right of $z = 0$ is the mirror image of the portion to the left of $z = 0$. Thus the value $z = 0$ divides the curve and the total probability of 1.0 in half. Hence the probability that z will be less than 0 is the same probability that z will be greater than 0, namely 0.5.

There is no simple formula that gives the areas under the normal curve for various intervals, so such areas are generally given in the form of tables. A table is given in the Appendix (page 180). But before one turns to tables one can get a good idea of these probabilities by memorizing certain key values.

We divide the z-axis into 8 key intervals. The probability assigned to each of these intervals is as follows:

Interval			Probability
$-\infty$	to	-2	0.02
-2	to	-1	0.14
-1	to	$-1/2$	0.15
$-1/2$	to	0	0.19
0	to	$1/2$	0.19
$1/2$	to	1	0.15
1	to	2	0.14
2	to	∞	0.02
Total	$-\infty$ to	∞	1.00

This table may be shown graphically as

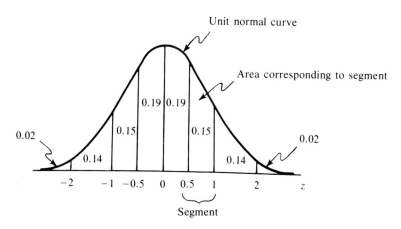

Unit normal curve

Area corresponding to segment

Segment

From this table or picture we can build up various probabilities. Certain relationships hold among areas and hence among probabilities. As long as areas don't overlap we can add them. That is, as long as intervals don't overlap, we can add the corresponding probabilities.

Exercises

1. The probability of z being less than 0.5 is
   (a) $P(0.5 < z < 1.0)$
   (b) $P(-1.0 < z < 0.5)$
   (c) $P(-\infty < z < 0.5)$
   (d) $P(0.5 < z < \infty)$
2. For a standard normal distribution, the probability that z will occur below 2 is
   (a) 0.48 (b) 0.99 (c) 0.98 (d) 0.49
3. A particular normal distribution is determined by the values of its
   (a) mode and mean
   (b) mean and standard deviation
   (c) median and mode
   (d) median and mean
4. On the unit normal curve $P(-1.0 < z < 0) + P(0 < z < 0.5)$ corresponds to the shaded area in

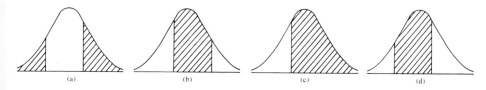

5. Given $P(-\infty < z < 0.5) = 0.69$, then $P(0.5 < z < \infty)$ is
   (a) 0.19   (c) 0.31
   (b) 0.69   (d) 0.50
6. Which is not correct for the standard normal distribution?
   (a) The total area under the curve is one.
   (b) The mean occurs at $z = 0$.
   (c) The probability of any given value of z is 0.5.
   (d) The curve is symmetric about the mean.
7. Since the standard normal curve is symmetrical about the mean $\mu = 0$, which would be correct?
   (a) $P(-\infty < z < \infty) = 0.5$
   (b) $P(0 < z < \infty) = 1.0$
   (c) $P(-\infty < z < 0) = P(0 < z < \infty)$
   (d) None of these
8. For a standard normal curve, the probability that z will fall below a given value K plus the probability that z will fall above the given value K is
   (a) $P(z < K) + P(z < \infty)$
   (b) $P(z > K) + P(z > 0)$
   (c) one
   (d) 0.50
9. For a standard normal distribution which of the following is correct
   (a) $P(-\infty < z < -1) = 0.84$
   (b) $P(-\infty < z < 1) = 0.84$
   (c) $P(-1 < z < 1) = 0.84$
   (d) $P(1 < z < \infty) = 0.84$
10. For a standard normal distribution which is correct?
    (a) $P(-1 < z < \infty) = 0.16$
    (b) $P(-\infty < z < 2) = 0.02$
    (c) $P(-\infty < z < -2) = 0.98$
    (d) $P(-\infty < z < -2) = 0.02$
11. Using the figure given for the standard normal curve, find the:
    (a) Probability that a z-score will be greater than 1
    (b) Probability that a z-score will be between 0.5 and 2
    (c) Probability that a z-score will be greater than -0.5
    (d) Probability that a z-score will be between -1 and 2
    (e) Find the z-score such that the probability of a larger value is 0.02
    (f) Find the z-score such that the probability of a larger value is 0.84
    (g) Suppose that the z-score lies between $-b$ and $-b$. Find the value of b such that the probability is 0.96
12. Given the correspondence between x and z as shown by:

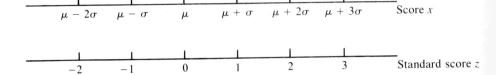

(a) What is the formula relating x and z?

(b) Suppose the mean $\mu = 75$ and the standard deviation $\sigma = 10$, for a normal population. What percentage of all scores lie between 70 and 80?

Hint. Convert x scores to z scores. Then find area of segment between the two z scores.

13. The mean $\bar{x}$ of a sample of size n has a sampling distribution with mean μ and variance σ^2/n. The standardized mean is

$$z = \frac{\bar{x} - \mu}{\sigma} \sqrt{n}$$

Given the normal distribution with $\mu = 75$ and $\sigma = 10$. What is the probability that the mean $\bar{x}$ of a sample of $n = 25$ will differ from the population mean μ by less than one.

14. The length of adult snakes of a certain species are found to have a normal distribution with mean 14 in. and standard deviation 2 in.

(a) Sketch the normal curve.

(b) What fraction of these snakes are over 16 in. long? under 12 in? between 12 and 16 in.?

(c) What fraction of these snakes are between 14.4 in. and 18 in.? between 12.6 and 15 in.? between 12 and 12.6 in.?

Solution:

(a) To sketch this bell-shaped curve, first mark the mean and standard deviation on the x-axis, then find the height of the bell $\dfrac{0.4}{\sigma} = \dfrac{0.4}{2} = 0.2$ and plot this point above the mean, then mark points over the standard deviation about 3/5 the height of the bell, then draw in a bell-shaped curve through these points.

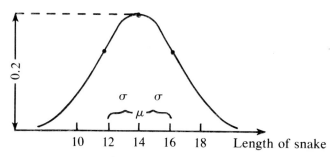

(b) 16 in is 2 in. over the mean, or one standard deviation above the mean. You should remember that .16 or approximately 1/6 of cases lie more than one standard deviation above the mean for a normal curve. Similarly 12 in. is one standard deviation below the mean, and 0.16 of cases occur below this level. Also 0.68 of cases lie within one standard deviation of the mean, i.e. between 12 and 16 in.

(c) To answer these questions, you need the Table for the Normal

Distribution as given in the Appendix (page 180). It is a good idea to shade in the area required for each question. The fraction *between* 14.4 *and* 18 *inches* is shown by the shaded area

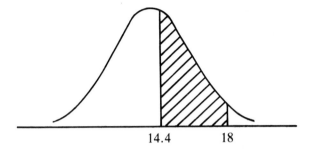

One can already guess that answer to be about 0.4, since a little less than half the area under the curve is shaded in. Now calculate the z-values, the number of standard deviations out from the mean for 14.4 in. and 18 in. We see that 14.4 in. is 0.4 in. above the mean or $z =$ $\dfrac{0.4 \text{ in.}}{2 \text{ in.}} = 0.2$ standard deviations above the mean. Similarly 18 in. is z $= 2$ standard deviations above the mean. The Table gives the area between $z = 0$ (the mean) and each value of z in the table: for $z = 0.2$, the area of 0.0793 given in the table represents this area:

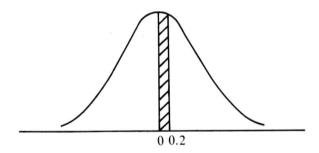

while for $z = 2$, the value of 0.4772 represents this area:

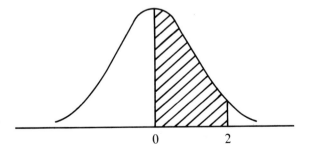

We want the area between $z = 0.2$ and $z = 2$

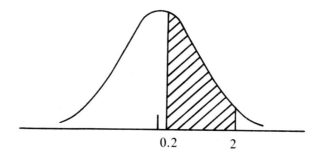

which is clearly the second figure *minus* the first figure. Answer = $0.4772 - 0.0793 = 0.3979$ of snakes have lengths between 14.4 in. and 18 in.

Between 12.6 and 15 inches:

$$\begin{aligned} \text{Total area} &= (\text{Area from } z = -0.7 \text{ to } 0) \\ &\quad + (\text{Area from } z = 0 \text{ to } 0.5) \\ &= (\text{Area from } z = 0 \text{ to } 0.7) \\ &\quad + (\text{Area from } z = 0 \text{ to } 0.5) \\ &= 0.1915 + 0.2580 = 0.4495 \end{aligned}$$

Between 12 and 12.6 inches:

$$\begin{aligned} \text{Total area} &= \text{Area from } z = -1 \text{ to } -0.7 \\ &= \text{Area from } z = 0.7 \text{ to } 1 \\ &= (\text{Area from } z = 0 \text{ to } 1) - (\text{Area from } z = 0 \text{ to } 0.7) \\ &= 0.3413 - 0.2580 = 0.8333 \end{aligned}$$

15. Company X makes pound-boxes of butter, but variations (errors) naturally occur from box to box. Suppose to be on the safe side, the management sets the machine to produce boxes of average net weight a little above 1 lb., say at 1.01 lb. The variability makes the distribution of new weights in the boxes a normal curve with mean 1.01 lb. and standard deviation 0.02 lb.

    (a) Sketch the density function and give its formula.

    (b) What fraction of boxes exceed 1.03 lb. in weight? are less than 0.99 lb. in weight? are between 0.99 and 1.03 lb.?

    (c) What fraction of boxes are underweight, i.e., less than 1.00 lb.?

16. Company Y is a shoddy outfit also making pound boxes of butter. Its machines are less accurate, so that the standard deviation is 0.05 lb. To make matters worse, the management cheats a little and sets the machine to average only 0.98 lb.

    (a) Sketch the density function, assumed normal.

    (b) 0.16 of the boxes exceed _____ in net weight, while 0.16 of the boxes are below _____ in weight, and the remaining 0.68 are between _____ and _____ in weight.

    (c) What fraction of the boxes are underweight?

17. The College Entrance Examination Board test scores are scaled to ap-

proximate a normal distribution with mean μ = 500 and standard deviation σ = 100.

    (a) Sketch the density function and give its formula

    (b) 0.16 of the students *taking the exam* score over _____, while 0.16 of the students score below _____, and the remaining 0.68 of students score between _____ and _____.

    (c) What fraction of students taking the exam score between 450 and 650? over 650?

  18. Which of the following would have an approximately normal distribution? If possible, sketch the distribution (normal or non-normal)

    (a) adult male's heights

    (b) the distribution of family incomes in the U.S.

    (c) heights of all people in the U.S.

    (d) estimates of the length of an object by a large group of people all using the same ruler in turn

    (e) daily production of a certain factory

  19. The area under the normal curve is

    (a) equal to 0.95

    (b) between plus and minus 2 standard deviations

    (c) one

    (d) one hundred

  20. For a normal distribution with a mean of 60 and a standard deviation of 10, what is the chance that a score is greater than 60?

    (a) 0.25   (c) 0.50

    (b) 0.196   (c) 70

CHAPTER 8

EXPECTATION

Introduction

One of the most useful concepts in statistics is that of expected value or expectation, denoted by the symbol E. In words, the symbol E is an operator that stands for any of the equivalent expressions given here:

Expectation of
Expected value of
True mean of
Population mean of
E of

For example, instead of saying that μ is the population mean of the score x, we can merely write

$$\mu = Ex,$$

which in words is μ is equal to E of x, or just Ex. The expected value can always be thought of a population average of the quantity that follows the symbol E. The symbol E is a convenient and useful shorthand notation for a concept which is used over and over again in statistics.

Usually we use the symbol μ alone, but to explicitly show that μ is the mean of x we can attach x to μ as a subscript. Thus,

$$\mu_x = Ex.$$

Properties of expectation

Now let us develop some useful properties connected with the expected values of different kinds of scores. The simplest kind of score is one capable of taking on only a single value c. Such a score that doesn't change at all, of course, is a constant. For example $c = 100$ would correspond to the case of giving everyone in a class 100 percent regardless of his performance on the examination. Since the score will always equal c, it will clearly be equal to c on the average, so its expected value must be c. We express this property as

$$Ec = c$$

and say that the expected value of a constant score c is equal to c.

69

Next let x be any score, and c be a constant. We can obtain a new score by adding c to the value of x. For example we might add 50 points to each score obtained on an examination. The new score would be given by $x + c$. The expectation of the new score $x + c$ is given by

$$E(x + c) = Ex + c$$

and we say that the expected value of the new score $x + c$ is equal to the expected value of the old score x plus the constant c. In words, when a constant c is added to a score, the same constant c is added to its expectation. For example, if each score x is increased by 50 points, then the class average will necessarily be increased by 50 points also.

Let us next find out what happens when we multiply each score x by a constant c. For example, suppose we multiply each score by 10. We would find that the new class average is then 10 times that of the old class average. In symbols we have

$$E(cx) = cEx$$

which in words is that the expected value of cx is equal to c times the expected value of x.

Finally we come to one of the most useful properties of expectation, namely the addition property. Let each student obtain two scores x and y, and let us obtain a new score by adding them together. The new score is then the sum $x + y$. The class average of the new score is then equal to the sum of the class averages of the old scores. That is,

$$E(x + y) = Ex + Ey$$

or, in words, the expected value of the sum of two scores is the sum of the expected values of each score.

It is easy to extend this addition property to more than two kinds of scores. For example, each student in a class may receive a score x in math, y in English, and z in science. The students total score in the three subjects is $x + y + z$. The class average of the total score $x + y + z$ is the sum of the averages Ex, Ey, Ez of the individual subjects, that is

$$E(x + y + z) = Ex + Ey + Ez.$$

More generally, if $x_1, x_2, \ldots, x_n$ are any n types of scores, then the expected value of their sum is equal to the sum of their expected values, which in symbols is

$$E(x_1 + x_2 + \ldots + x_n) = Ex_1 + Ex_2 + \ldots + Ex_n.$$

Variance

The expected value of a score x is equal to the true mean (or population mean) μ of the score, that is $Ex = \mu$. We recall that the population variance of the score x is a true measure (or population measure) of the extent to which the population distribution of scores is dispersed away from the true mean μ. More specifically, the population variance of the score x is the expected value of $(x - \mu)^2$, that is

$$\text{var } x = E(x - \mu)^2$$

In words, the population variance of a score is the expectation of the square of the difference between the score and its expectation. Before, we have introduced the symbol σ^2 for the population variance, so

$$\sigma^2 = \text{var } x = E(x - \mu)^2.$$

However the symbol var x is more explicit, as it explicitly shows that we are dealing with the score x.

Variance has properties analogous to the properties of expectation. The first property is that for any constant c,

$$\text{var } c = 0.$$

The reason for this property is as follows. If a score is always equal to a constant, then the constant will be the expectation of the score. That is, if $x = c$ then $\mu = c$ also. Therefore the score never differs from its expectation, and hence the difference $x - \mu$ is always $c - c$ or 0. The square of zero is zero, and so is its expectation.

The next property is that for any constant c

$$\text{var } (x + c) = \text{var } x.$$

We can obtain this result as follows. If we add the constant c to the score x, the expectation of x is also increased by c. It follows that the difference $x - \mu$ is unchanged, so that the variance is also unchanged.

If c is any constant, then

$$\text{var}(cx) = c^2 \text{ var } x.$$

We justify this property as follows. Multiplying the score x by c will also multiply its expectation μ by c. Hence the difference is $cx - c\mu = c(x - \mu)$. That is, the new difference is c times as large as before. The square of the new difference is $c^2(x - \mu)^2$ which is c^2 times as large as before. It follows that

$$\text{var } cx = E[c^2(x - \mu)^2] = c^2E(x - \mu)^2 = c^2\text{var } x,$$

that is, the variance of cx is c^2 times the variance of x. In the special case when $c = -1$ we have

$$\text{var } (-x) = (-1)^2 \text{ var } x = \text{var } x.$$

This property shows that multiplication of a score by a constant results in the multiplication of the variance of the score by the square of the constant. This is because the variance is the expectation of a squared quantity. If we convert from feet to inches, each score is multiplied by 12, but the variance is multiplied by

$(12)^2 = 144$. As a result it is usually more convenient to use a spread parameter that changes by the same scale factor, rather than by the square factor. As we have seen, such a parameter is the standard deviation σ, defined as the positive square root of the variance. Usually we use the symbol σ alone, but to explicitly show that σ is the standard deviation of the score x we can attach x to σ as a subscript. Thus

$$\sigma_x = \sqrt{\operatorname{var} x}$$

or equivalently

$$\sigma_x^2 = \operatorname{var} x.$$

The foregoing property for variance becomes

$$\sigma_{cx} = c\sigma_x$$

which says that the standard deviation of a constant times a score is equal to the constant times the standard deviation of the score. This property follows from

$$\sigma_{cx} = \sqrt{\operatorname{var}(cx)} = \sqrt{c^2 \operatorname{var} x} = c\sqrt{\operatorname{var} x} = c\sigma_x.$$

Thus the standard deviation of a score in inches is just 12 times the standard deviation of the score in feet.

We now come to the addition property. By analogy with the addition property of expectations, we would hope that there would be an addition property for variances that would say that the variance of the sum of two scores is the sum of their variances. Such an addition property is true provided that the two scores are *uncorrelated*. We will take up correlation later, but for the moment we need to know only that independent scores are always uncorrelated. Thus for two independent scores x and y we have the addition property for variances given by

$$\operatorname{var}(x + y) = \operatorname{var} x + \operatorname{var} y.$$

In words, the variance of the sum of two independent scores x and y is equal to the sum of the variance of x and the variance of y. More generally, the variance of the sum of n independent scores is equal to the sum of their variances:

$$\operatorname{var}(x_1 + x_2 + \ldots + x_n) = \operatorname{var} x_1 + \operatorname{var} x_2 + \ldots + \operatorname{var} x_n$$

Random sample

We recall that a random sample of size n is a set of n independent draws $x_1, x_2, \ldots, x_n$ from a given population. If the population mean is μ and the population variance is σ^2, then each of these draws has μ as its expected value and σ^2 as its variance. That is,

$$
\begin{aligned}
Ex_1 &= \mu, & \operatorname{var} x_1 &= \sigma^2 \\
Ex_2 &= \mu, & \operatorname{var} x_2 &= \sigma^2 \\
&\cdots & &\cdots \\
Ex_n &= \mu, & \operatorname{var} x_n &= \sigma^2.
\end{aligned}
$$

The sample mean is

$$\bar{x} = \frac{x_1 + x_2 + \ldots + x_n}{n}$$

which we can write as

$$\bar{x} = \frac{1}{n} \Sigma x$$

The expected value of the sample mean is

$$E\bar{x} = E\left(\frac{1}{n} \Sigma x\right) = \frac{1}{n} E(\Sigma x) = \frac{1}{n} E(x_1 + x_2 + \ldots + x_n)$$

$$= \frac{1}{n} (Ex_1 + Ex_2 + \ldots + Ex_n) = \frac{1}{n} (\mu + \mu + \ldots + \mu) = \frac{1}{n} (n\mu)$$

which finally gives

$$E\bar{x} = \mu.$$

In words, *the expected value of the sample mean is equal to the population mean.*
    The variance of the sample mean is

$$\text{var } \bar{x} = \text{var} \left(\frac{1}{n} \Sigma x\right) = \frac{1}{n^2} \text{var } (\Sigma x)$$

$$= \frac{1}{n^2} \text{var } (x_1 + x_2 + \ldots + x_n)$$

$$= \frac{1}{n^2} (\text{var } x_1 + \text{var } x_2 + \ldots + \text{var } x_n)$$

$$= \frac{1}{n^2} (\sigma^2 + \sigma^2 + \ldots + \sigma^2)$$

$$= \frac{1}{n^2} n\sigma^2$$

which finally gives

$$\text{var } \bar{x} = \frac{\sigma^2}{n}.$$

In words, *the variance of the sample mean is equal to the population variance
divided by the sample size.* Taking square roots we have

$$\sigma_{\bar{x}} = \frac{\sigma}{\sqrt{n}}.$$

In words, *the standard deviation of the sample mean is equal to the population
standard deviation divided by the square root of the sample size.*

Sample variance

We recall that the sample variance is defined as

$$s^2 = \frac{1}{n-1} \Sigma(x - \bar{x})^2.$$

If we define a new score y as $y = x - \mu$, then clearly

$$x - \bar{x} = y - \bar{y}$$

so in terms of the new score the sample variance is

$$s^2 = \frac{1}{n-1} \Sigma(y - \bar{y})^2.$$

That is, we can substitute y for x and $\bar{y}$ for $\bar{x}$ in the formula for s^2. We also recall the alternate formula for the sample variance given by (see page 10)

$$s^2 = \frac{1}{n-1} (\Sigma x^2 - n\bar{x}^2).$$

Because we can substitute y for x and $\bar{y}$ for $\bar{x}$, we obtain the equivalent formula

$$s^2 = \frac{1}{n-1} (\Sigma y^2 - n\bar{y}^2).$$

Taking expectations, we have

$$Es^2 = \frac{1}{n-1} E(\Sigma y^2 - n\bar{y}^2)$$

which is

$$Es^2 = \frac{1}{n-1} [E(\Sigma y^2) - nE\bar{y}^2].$$

Let us now look at the terms in the square brackets on the right. The first term can be written as

$$
\begin{aligned}
E(\Sigma y^2) &= E(y_1^2 + y_2^2 + \ldots + y_n^2) \\
&= Ey_1^2 + Ey_2^2 + \ldots + Ey_n^2 \\
&= E(x_1 - \mu)^2 + E(x_2 - \mu)^2 + \ldots + E(x_n - \mu)^2 \\
&= \text{var } x_1 + \text{var } x_2 + \ldots + \text{var } x_n \\
&= \sigma^2 + \sigma^2 + \ldots + \sigma^2 = n\sigma^2.
\end{aligned}
$$

Here we have used the fact that $y = x - \mu$. The second term is

$$-nE\bar{y}^2 = -nE(\bar{x} - \mu)^2 = -n \text{ var } \bar{x}.$$

Here we have used the fact that $\bar{y} = \bar{x} - \mu$. From the last section we know

$$\text{var } \bar{x} = \frac{\sigma^2}{n}.$$

Thus the second term is

$$-nE\bar{y}^2 = -n\frac{\sigma^2}{n} = -\sigma^2.$$

The equation for Es^2 becomes

$$Es^2 = \frac{1}{n-1}[n\sigma^2 - \sigma^2] = \frac{1}{n-1}(n-1)\sigma^2 = \sigma^2.$$

In words, *the expectation of the sample variance is equal to the population variance.*

Unbiased estimates

We recall that a statistic is a quantity computed from the data. For example the sample mean $\bar{x}$ is found by summing the sample observations and then dividing the sum by the number n of observations. The sample mean $\bar{x}$ can be used as an estimate of the population mean μ. Because the expected value of $\bar{x}$ is equal to μ we say that $\bar{x}$ is an unbiased estimate of μ.

Generally, we say that *an estimate is unbiased if its expected value is equal to the population parameter that we are trying to estimate.*

As another example we compute the statistic s^2 to estimate the population variance σ^2. As we have seen in the preceding section, the expected value of s^2 is equal to σ^2. Thus s^2 is an unbiased estimate of the population variance σ^2. We recall that s^2 is found by summing the squared deviations of x from the sample mean $\bar{x}$ and then dividing the result by $n-1$. If instead we had divided by n, we would not have obtained an unbiased estimate.

In some cases we may know the population mean μ. In these cases we can form the estimate of the population variance given by

$$\frac{1}{n}\Sigma(x-\mu)^2.$$

Because

$$E\left[\frac{1}{n}\Sigma(x-\mu)^2\right] = \frac{1}{n}\Sigma E(x-\mu)^2$$

$$= \frac{1}{n}\Sigma\sigma^2 = \frac{1}{n}(\sigma^2 + \sigma^2 + \ldots + \sigma^2)$$

$$= \frac{1}{n}(n\sigma^2) = \sigma^2$$

we see that this estimate is unbiased.

The important points to remember are these: If the deviations are measured about the sample mean $\bar{x}$, then we divide the sum of squared deviations by $n-1$ in order to obtain an unbiased estimate of σ^2. On the other hand if the deviations are measured about the population mean μ, then we divide the sum of squared deviations by n in order to obtain an unbiased estimate of σ^2.

Exercises

1. Let the fixed monthly salary of a salesman be 500, and let x be the random amount he earns through commissions during a month. If the expected value of his monthly commissions is 750, what is the expected monthly income?

2. If a gambler on straight bets expects to earn 100, and on side bets 50, what is his expected total earnings?

3. Because a class did poorly on an exam, the teacher doubled everybody's grade. If the old class average is 37 what is the new class average?

4. Show that $E(x - y) = Ex - Ey$. Hint: $x - y = x + (-1)y$.

5. If c and d are constants, show that $E(cx + dy) = cEx + dEy$.

6. Suppose that each of the three scores x, y, z has the same expectation μ. Show that $E(x + y + x) = 3\mu$.

7. Suppose that each of the n scores $x_1, x_2, \ldots, x_n$ has the same expectation μ. Let $\bar{x}$ be the arithmetic mean $(x_1 + x_2 + \ldots + x_n)/n$. Show that $E\bar{x} = \mu$.

8. If the physical education scores x have variance 25 and the science scores y have variance 16 (and the scores are independent) what is the variance of the score $x + y$.

9. If a score is multiplied by a negative constant, then is the resulting standard deviation negative?

10. If x and y are independent, x has a standard deviation of 6 and y has a standard deviation of 8, then what is the standard deviation of their sum?

11. If x and y are independent and each has the same standard deviation σ, then what is the standard deviation of the sum $x + y$?

12. If x and y are independent and c and d are constants, show that

$$\text{var } (cx + dy) = c^2 \text{ var } x + d^2 \text{ var } y.$$

13. If x and y are independent, show that

$$\text{var } (x - y) = \text{var } x + \text{var } y.$$

14. Suppose that each of the three independent scores x, y, z has the same variance σ^2. Show that

$$\text{var } (x + y + z) = 3\sigma^2.$$

15. Suppose that each of the n independent scores $x_1, x_2, \ldots, x_n$ has the same variance σ^2. Let $\bar{x}$ be their arithmetic mean $(x_1 + x_2 + \ldots + x_n)/n$. Show that var $\bar{x}$ $= \sigma^2/n$.

16. If the expected value of x is 5 then the expected value of $-x$ is
(a) 0.0   (b) 0.5   (c) 0.25   (d) -5

17. If the variance of x is 5 then the variance of $-x$ is
(a) -5 (b) $\sqrt{5}$ (c) 5   (d) 25

18. If the expected value of x is 5 and the expected value of y is 10 then the expected value of $x - y$ is
    (a) 15 (b) 10 (c) 5 (d) -5

19. If the standard deviation of x is 3, if the standard deviation of y is 4, and if x and y are independent, then the standard deviation of $x + y$ is
    (a) 3 (b) 4 (c) 5 (d) 7

20. If the expected value of x is 50 then the expected value of $\bar{x}$ is
    (a) 0 (b) 50 (c) $50n$ (d) $\dfrac{50}{\sqrt{n}}$

21. If the standard deviation of x is 5 then the standard deviation of $\bar{x}$ is
    (a) 5 (b) $5n$ (c) $\dfrac{5}{\sqrt{n}}$ (d) $\dfrac{5}{n}$

22. If the standard deviation of x is 5 then the standard deviation of $10x$ is
    (a) 5 (b) 25 (c) 50 (d) 250

23. If the variance of x is one, then the variance of $10x$ is
    (a) 1 (b) 10 (c) 100 (d) $-\sqrt{100}$

24. If x and y are independent, and if x has standard deviation 4 and y has standard deviation 3, the $x - y$ has standard deviation
    (a) 1 (b) $\sqrt{7}$ (c) 5 (d) 7

25. If the variance of x is σ^2 then the standard deviation of $\bar{x}$ is
    (a) σ^2/n (b) $n\sigma^2$ (c) $\sigma/\sqrt{n}$ (d) σ/n

26. Given the sets of numbers 2, 5, 8, 11, 14 and 2, 8, 14, find (a) the mean of each set, (b) the variance of each set, (c) the mean of the combined or "pooled" sets, (d) the variance of the combined or pooled sets.

27. A student received a grade of 84 on a final examination in mathematics for which the mean grade was 76 and the standard deviation was 10. On the final examination in physics for which the mean grade was 82 and the standard deviation was 16, he received a grade of 90. In which subject was his relative standing higher?

CHAPTER 9

ENTROPY

Definition of entropy

"Something there is that doesn't love a wall, That wants it down," wrote Robert Frost. Rocks weather and crumble, iron rusts, people grow old. Heat always flows spontaneously from a hotter to a colder body; fires grow cold. Gas always seeps spontaneously through an opening from a region of high pressure to a region of low pressure; tires go flat. Liquids left by themselves always tend to mix; spilled milk can never be regained. Work may be dissipated completely into heat, but this heat cannot be converted entirely back into work; cars run out of gasoline.

These are all examples of irreversible processes that take place naturally in only one direction, and by their one-sidedness express the direction of time. If we have two snapshots of a house that is not maintained, and if one snapshot was taken ten years ago and the other today, and if there were no dates on the snapshots, we could still tell which is which. The same is true for two snapshots of a person, one taken years ago and one today; we can tell them apart.

As something runs down, grows older, falls apart, becomes dissipated, and attains the multitude of other bad effects brought on by time, then we say that the thing is gaining *entropy*. In the end, a wall will be reduced to nothing more than dust; the wall has then attained its state of *maximum entropy*.

We do not know the total or absolute value of the entropy of a body; we can only measure changes in the entropy. It is an empirical fact (that is, a fact that is observed in nature, but not a fact that can be established by some theoretical argument) that all natural processes are accompanied by an increase in entropy with time. Everywhere then entropy is on the increase as time steals on. This observational fact is summed up in the so-called second law of thermodynamics which says: *The entropy of the universe tends to a maximum.* In other words, time is a thief.

The reader must not be discouraged if he does not yet clearly understand what entropy is. We cannot feel it like temperature or see its effects as we can see the effects of heat or pressure. Our knowledge is necessarily roundabout, and our conception of it will perhaps always be a little vague. There is nothing else like entropy in the universe; there is nothing to which entropy can be accurately compared. All other variables can be increased or decreased; but entropy on the whole always increases. Entropy is a one-way street. It can only be decreased temporarily and in a restricted region, and then only at the expense of a greater increase elsewhere. We can fill up the car again but it will run out of gasoline soon again, and we can never make anymore gasoline to replace the gasoline used up. The sun did make the gasoline over millions of years during the past eons, but the sun itself is running down.

The general increase in entropy is inexorable and unpreventable; it represents the passage of time. Entropy is the physical variable that unequivocally marks the universe as older today than it was yesterday. Entropy is time's arrow. As age creeps upon us unaware we scarcely observe the signs until they have become conspicuous. Signs of old age are a decrease in adaptability and a loss of available energy. With old age there is a lessened variety, a decreased transformability, and a trend to uniform monotomy.

The ancients understood entropy. The Egyptians built the pyramids: "Time defies all things, but the pyramids defy time." But they know, and we know, that this statement is not true. Time will lay low the pyramids. Time marches on.

Simple examples

Perhaps the simplest possible example is that of a universe which has only two locations, say A and B, and two atoms. Initially suppose that both atoms are in location A:

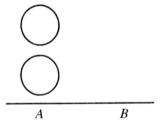

Since these two atoms make up the entire population, we can say that the probability of finding an atom in location A is one and the probability of finding an atom in location B is zero; that is, we have the initial probabilities

$$P(A) = 1$$
$$P(B) = 0.$$

The piling up of atoms in location A represents a wall, and according to Robert Frost entropy doesn't love a wall; entropy wants the wall down. Thus with the passage of time we would expect our universe to look as follows:

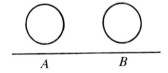

In other words, the effect of time would be to move one of the two atoms that were in location A to location B. Because entropy doesn't love a wall it would not leave both atoms in A, nor would it move both atoms to B thereby producing another wall:

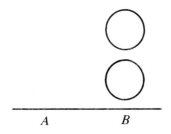

The passage of time would only result in one atom in A and one atom in B so we have the final probabilities:

$$P(A) = 0.5$$
$$P(B) = 0.5.$$

In terms of the probability distribution of an universe (or population) we can actually define a *measure of entropy* which we call H. The entropy H in the case of two locations (or possibilities) is defined as

$$H = -P(A) \log P(A) - P(B) \log P(B).$$

In the case of three possibilities A, B, C the entropy is defined as

$$H = -P(A) \log P(A) - P(B) \log P(B) - P(C) \log P(C).$$

In these expressions $\log P(A)$ represents the logarithm of $P(A)$, likewise $\log P(B)$ represents the logarithm of $P(B)$, and similarly for $\log P(C)$. It is often convenient to use 2 as the base of the logarithms. The use of 2 as the logarithmic base has advantages in the case of distributions whose probabilities are multiples of 1/2. If we use the base 2, then we have

$$\begin{aligned}
\log 1 &= 0 \\
\log (1/2) &= -\log 2 = -1 \\
\log (1/4) &= \log (1/2)^2 = 2 \log (1/2) = -2 \\
\log (1/8) &= -3 \\
\log (1/16) &= -4 \\
\log (1/2^n) &= -n.
\end{aligned}$$

Let us now return to our simple example. The initial universe has probabilities $P(A) = 1$, $P(B) = 0$ so its entropy is

$$H = -1(\log 1) - 0(\log 0) = -1(\log 1) = 0.$$

The final universe has probabilities $P(A) = 1/2$, $P(B) = 1/2$ so its entropy is

$$H = -1/2 \log(1/2) - 1/2 \log(1/2) = 1/2 + 1/2 = 1$$

Thus with the passage of time entropy has increased from 0 to 1. The final universe is one of maximum entropy, which in this case is one.

Let us now consider another example, one with three locations A, B, C. We can think of this universe as being built up of two simpler universes, namely a sub-universe with locations A, B and a sub-universe with locations B, C. If there are two atoms in each sub-universe, then the states of maximum entropy for each of the two sub-universes would be one atom in each position:

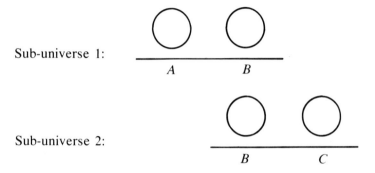

Sub-universe 1:

Sub-universe 2:

As a result the state of maximum entropy of the entire universe would be one atom in location A, two atoms in location B, and one atom in location C:

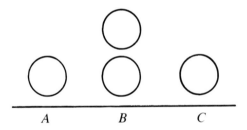

We see that there is a wall at location B, but it is a low wall and it is in the center. It is in the center because an atom can reach the center location B from either location A or location C, whereas the end locations A or C can only be reached from one location, namely the center location B. In other words the central location is the most mediocre one. The final probabilities are

$$P(A) = 1/4, \; P(B) = 1/2, \; P(C) = 1/4$$

so the final (or maximum) entropy is

$$H = -1/4 \log 1/4 - 1/2 \log 1/2 - 1/4 \log 1/4$$
$$= 1/2 + 1/2 + 1/2 = 1.5$$

The lowest possible entropy would occur with a high wall such as in the universe:

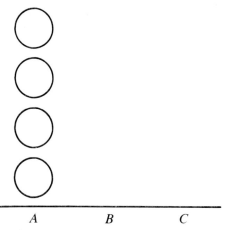

The entropy for this universe is

$$H = -P(A) \log P(A) = -1 \log 1 = 0$$

The second law of thermodynamics says that the universe cannot stand the ravages of time; in the end only a final or maximum-entropy universe

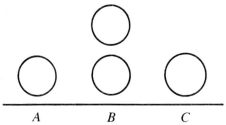

would result. The entropy would have increased from the minimum of 0 to the maximum of 1.5.

Entropy as a measure of uncertainty

Suppose we come to a branch point where one of two possible results must occur:

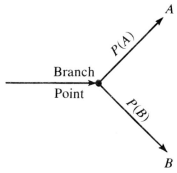

Let these two cases be *A* and *B* with probabilities $P(A)$ and $P(B)$ respectively. The sum of these probabilities $P(A) + P(B)$ is one, as the two cases are exhaustive.

The quantity

$$I(A) = -\log P(A)$$

is called the amount of *self-information* associated with the event *A*. If we use logarithms to the base 2, then the unit of the amount of information is called a bit. For example, if the events *A* and *B* are equally likely, so $P(A) = 1/2$, then the self-information of event *A* is

$$I(A) = -\log 1/2 = 1 \text{ bit}$$

so the occurrence of event *A* conveys to us 1 bit of information. That is, the occurrence of one of two equally likely events conveys one unit (i.e. one bit) of information. On the other hand suppose the event *A* was certain, that is $P(A) = 1$. Then its self-information is

$$I(A) = -\log 1 = 0.$$

That is, the occurrence of a sure thing conveys no information. As another example, suppose that the event A has an only 1 chance in 4 to happen; that is, suppose $P(A) = 0.25$. Then its self-information is

$$I(A) = -\log 1/4 = 2 \text{ bits}.$$

In other words, the occurrence of an event with one chance in four conveys to us 2 bits of information. But such an event is equivalent to two independent events each of which has one chance out of two. That is, the following two diagrams are equivalent.

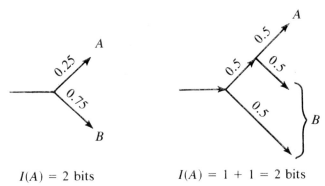

$I(A) = 2 \text{ bits}$ $I(A) = 1 + 1 = 2 \text{ bits}$

Exercises

1. Suppose each fork represents a $50 - 50$ choice. The following map shows the way home:

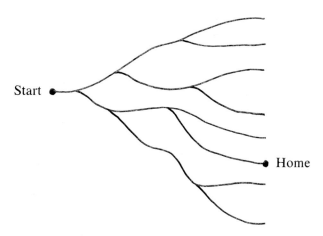

The correct message to find the way home is (where L = left, R = right)

   (a) *LLR*   (b) *LRL*   (c) *RLR*   (d) *RRL*

Because the correct message contains (and requires) 3 symbols (each of which is 50-50) we say that the information content of the message is 3 bits.

   2. How many bits are required to find the way home from the map

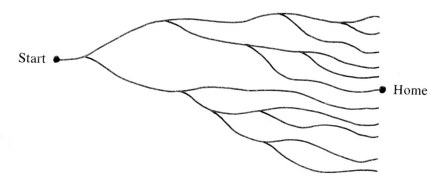

CHAPTER 10

THE UNCERTAINTY PRINCIPLE

Basic idea of the uncertainty principle

A ship at sea obtains a fix on one lighthouse. The direction of the lighthouse is certain (i.e., its variance is zero) but the position of the ship is uncertain (i.e., its variance is infinite).

Lighthouse

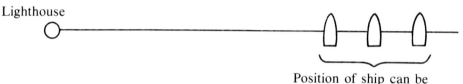

Position of ship can be anywhere on line of direction

A ship at sea obtains a fix on two lighthouses. There is now a spread in directions (i.e., the variance of the direction is positive and no longer zero) and now the position of the ship is certain (i.e., the variance is zero).

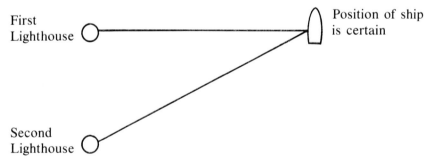

First Lighthouse

Position of ship is certain

Second Lighthouse

The variables *position of ship* and *lighthouse direction* are intimately related by the fact that a small variance in one is necessarily associated with a large variance in the other. Two variables related in this way are said to satisfy the *uncertainty principle*. Two such variables may be referred to as *dual variables*.

As another example, consider a pleasant, friendly, good-looking politician. If he is vague on the issues (i.e. issue variance large) he will offend the beliefs of few people and there will be general agreement on his charisma (i.e. charisma variance small). If he is specific on the issues (i.e. issue variance small) he will offend the beliefs on a large number of people and polarize them and as a result there will be pronounced disagreement on his charisma (i.e. charisma variance large). Thus we see that *issues* and *charisma* represent two variables that satisfy the uncertainty principle.

Let us consider the position of a candidate in the polls and his momentum due to his aggressive campaigning. If he has a well-defined momentum (i.e. momentum variance small), then that momentum will give an impetus and will affect undecided people, so his position in the polls becomes an uncertain indicator of the final result (i.e. position variance large). On the other hand, if his momentum is unclear (i.e. momentum variance large) it means that he is having little effect on the unknown quantities in the upcoming election, and as a result his position in the polls becomes a quite certain indicator of the final result (i.e. position variance small). In summary, we say that *position* and *momentum* satisfy the uncertainty relation.

We have said that two variables satisfy the uncertainty principle if a small variance in either one implies a large variance in the other one. However this statement does not necessarily operate in the reverse direction; that is, a large variance in either one does not necessarily mean a small variance in the other one. For example, a politician can have both a large variance in issues and a large variance in charisma. All the uncertainty principle states is that if one variance is small the other variance must be large; the uncertainty principle does not state that both variances cannot be large. In brief, the uncertainty principle states that both variances cannot be small. This point is illustrated in the diagrams for two variables x and y that satisfy the uncertainty principle:

Variance of x

		Small	Large
Variance of y	Small	Uncertainty principle says this is not possible	OK
	Large	OK	OK (Uniformly bad situation)

This diagram shows that the uniformly good situation (i.e. small-small) cannot be obtained, but that the uniformly bad situation (i.e. large-large) can surely be obtained. The impact is that the best we can do is represented by the two mixed situations (i.e. large-small and small-large) or by some compromise between the two mixed situations.

Application of the uncertainty principle

The uncertainty principle is often explained as follows. Random variables must be defined operationally, that is, they must be defined in terms of the statistical procedures through which they are measured. If we now analyze realistic statistical procedures of measurement, it turns out that a measurement will always perturb the system. In other words, there is a characteristic unavoidable interaction between the system and the measuring methods. As a result, the final measurement is not that of the original random variable itself but that of some perturbed version of it. So-called good measurement methods are ones that perturb the variable in such a way so as to reduce its variance. Since the perturbed variable then has a smaller variance than the original variable, it is felt that the result of the measurement is accurate and the measurement is successful. At this point in the argument the uncertainty principle is brought in.

Application of the uncertainty principle states that not only should the given variable be considered but also its dual variable. Any measurement method that perturbs the given variable so as to reduce its variance will necessarily perturb the dual variable so as to increase its variance. In other words, no measuring device can *simultaneously* take us into a uniformly good situation in which both variances are small. This fact should be taken into account in the evaluation of the measurement method.

For example, a person has both a position where he stands on an issue as well as inner vibrations (or vibes) on how he feels about the issue. If the issue is personally very close to him, he will have strong feelings that will make him differentiate very carefully between the various facets of different positions. As a result it is impossible for him to accept one definite fixed pat stand on the issue. On the other hand, if the issue is not important to him, his indefinite feelings about it makes it easy for him to say the issue is "yes" or "no", and thereby for him to take a clear definite stand. Thus we can say that a person's *stated position* and his *inner feelings* represent dual variables that satisfy the uncertainty principle. A poor man's feelings about hunger makes it difficult for him to take clean-cut positions on how much food everyday to give to his hungry animals, and the amount will vary from day to day. A rich man's feelings about hunger are so dispersed that he can give the same amount of food everyday to his animals without variation.

Generally people close to the problem have definite feelings about it and so it is difficult for them to take a "yes" or "no" position because they consider all the possibilities in between. People far removed from the problem have indefinite feelings and so more easily take a "yes" or "no" position.

Suppose that a person has indifferent feelings about television. His position about the programs is quite clear and definite. He might only watch football games and old movies when he has nothing else to do, and never watch prime-time productions. All at once, the Nielson Rating Service picks him as one of the 1200 people whose television viewing habits dictate what America will or will not see. Now his feelings about television become clear and definite, but his position on what TV programs to watch becomes indefinite and mixed, as he agonizes over whether he will watch this prime-time show or that one. Nielson is trying to find out this person's position on the various TV programs actually changed that person's variability. Beforehand the person's position had a small variance and his feelings a large variance, whereas afterwards the person's position had a large variance and his feelings a small variance.

As another example, suppose that when interviewers take a poll of public opinion, they find that everyone in a given city has almost the same opinion on a subject; that is, they find the variance on the opinion is very small for that city. This result disturbs the interviewers because the mean value of the opinion is not the result that fits in with their own preconceived opinions. As a result they take a new survey in which they pose the questions in such a way that it stirs up people's feelings on non-related as well as related subjects. In other words, in this loaded survey the questions deliberately bring forth definite and strong feelings on the part of the people being interviewed. The variance of the feelings of a person being interviewed is drastically reduced as a result of the type of question, and as a result the variance of his position greatly increases. The final result of the survey now has a much larger spread, and, even though the mean value of the opinion may not have changed, the variance around this mean has greatly increased. As a result the interviewers can now say that a large segment of the population on one side or the other side of the mean shares the preconceived opinion of the interviewers.

In summary, it might not be possible to change the mean value of a distribution, but with a judicious use of the uncertainty principle the variance can be changed. If the changed variance is smaller, the statement can be made that the mean value is a precise measure of the population, which is useful in situations where one wants to give credit to the mean. If the changed variance is larger, the statement can be made that the mean value is a poor representative of the population, which is useful in situations where it is desired to discredit the mean.

In some surveys the interviewers may be dissatisfied by the wide dispersion of a person's position. In such cases they would design questions that would disperse a person's feelings so that the person more readily takes a definite position. In other words, the questions have the effect of reducing the variance of the person's position at the expense of increasing the variance of the person's feelings.

The application of the *uncertainty principle* to measurement can be stated in this way: *Every measurement has the characteristic that as one quantity is measured with more precision (i.e. smaller variance) another related quantity is known with less precision (i.e. greater variance).*

Interference or interaction patterns

Suppose that we divide a first-grade class of school children into two groups: high-achievers and low-achievers. At the end of the eighth-grade the two groups might have grade distributions as shown in the diagram:

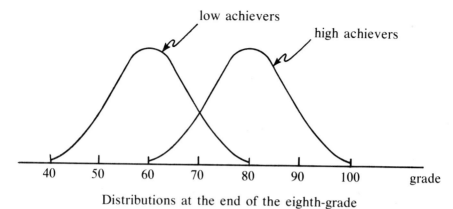

low achievers

high achievers

40   50   60   70   80   90   100   grade

Distributions at the end of the eighth-grade

The classification of first grade students into these two categories could be quite arbitrary such as by flipping a coin or could be by some sort of recognition system or achievement test.

On the other hand suppose that we did not separate the first grade class at all, and let them continue through school as one class. At the end of the eighth grade their grade distribution might be as shown by the solid curve in the diagram:

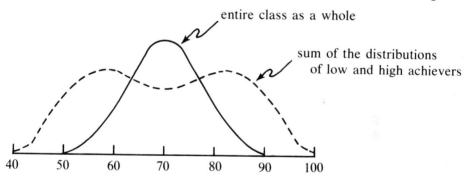

entire class as a whole

sum of the distributions
of low and high achievers

40   50   60   70   80   90   100

The important point we want to make is that generally the distribution for the entire class as a whole will not simply be the sum of the distributions of the split class. The reason is that there is an *interference* or *interaction effect* between high and low achievers when the entire class is together, and that there is no interference effect when the class has been separated into two distinct groups. In our diagrams we see that this interaction effect resulted in pulling more students in toward the mean; that is, when the class was not separated some low achievers did better and some high achievers did worse than thay would have done if they were separated.

Suppose now that a school administrator looks at the eighth grade of a class that has never been separated and picks out a student at random. Because the administrator believes in classification, he would think that either this child was a high achiever or a low achiever in the first grade, but not having the first grade records the administrator would not know which one the child was. In order to correct such a situation the administrator decides not to separate the current first grade class, but instead to put the spotlight on each high achiever in the first grade and continue to give these high achievers special attention through the eighth grade. As a result of this policy the grade distribution in the eighth grade would look like the dotted curve (i.e. the sum of the distributions of low and high achievers) in the above diagram, and not like the solid curve (i.e. the distribution of the entire class as a whole). That is, we no longer get the interaction distribution but instead one that shows no interaction. This result would be true despite the fact that the class was never actually separated into two groups.

We must conclude that when we look at the students as either low or high achievers the distribution of their grades is different than when we do not look. Perhaps it is the turning on of our spotlight on the student's achievement abilities that disturbs things. It must be that students are delicate and the tagging of them as either high or low achievers gives them a jolt that changes their progress. In fact we should expect their progress to be changed. By trying to watch the students we have changed their final distribution of grades.

You may think: "Don't make such an issue out of their status. Turn the special attention to high achievers down. As a result the students will not be disturbed as much. Surely by making the spotlight dimmer and dimmer, eventually the effect will be weak enough that it will have a negligible effect."

When we use this plan some students will go through school as high achievers and some as low achievers, but many will go through without any tagging at all. When we look at our distribution in the eighth grade we find these results. Those seen as high achievers will have a high-achiever distribution, and those seen as low achievers will have a low-achiever distribution, but those not seen as either will have the interference or interaction distribution of the whole.

In conclusion it is impossible to tag the students in some way in their education, and at the same time not disturb the pattern of the class as a whole. This result is an application of the *uncertainty principle*, which we can state in terms of this example as follows: *It is impossible to devise a method to classify students as to their achievements that will not at the same time disturb the students enough to change their interaction pattern.*

Exercises

1. Interpret the following in terms of the uncertainty principle: a public opinion poll is made more accurate by increasing the size of the sample. In fact the sample was made so large that it attracts everyone's attention to the question of interest. Speeches and discussions are held and various other methods are used to influence public opinion. Would the outcome of this sample represent fairly what the opinion would have been if no sample had been taken?

2. Suppose that in a public opinion poll the poll-takers move quietly from

place to place and unobtrusively strike up a casual conversation with various people. They ask the pertinent questions as though the answers were of no consequence. However because of this approach the poll is taken over a period of a year and they announce their results as they obtain them. Moreover many of the underlying factors have changed during that time. In terms of the uncertainty principle, discuss the accuracy of the poll.

3. True or false: The best way to determine the behavior pattern of an animal with the smallest possible variance is to lock the animal in a small cage at the zoo and let hundreds of observers view the animal. On the other hand, if one observer alone were to quietly observe the animal in the wild from a distance, only a large-variance estimate of the animal's behavior could be observed, because the animal's action would show greater variability and would be more difficult to see.

4. True or false: The best place to observe a tree is in a greenhouse, not a forest.

CHAPTER 11

DECISION MAKING

Null hypothesis, significance, and the level of significance

Statistics may be described as the science of making decisions from observations. Often we are faced with a decision in which the two alternatives are so equally balanced that we cannot make up our minds. For hours we will sit pondering first the one and then the other and all the time growing steadily unhappier. Indeed we become uncomfortably aware of the resemblance of our case to that of "Buridan's ass." The reference here is to the French philosopher, Jean Buridan, in the fourteenth century who stated the following: "If a hungry ass were placed exactly between two haystacks equal in every respect, the ass would starve to death, because there would be no motive why it should go to one rather than to the other haystack."

What Buridan didn't take into account was statistical theory, for his statement does not recognize the existence of a random factor. The ass is bound to turn his head in a random way so that one haystack comes into a better view, or to move his feet randomly so that one haystack becomes closer, and in the end he would be eating from the haystack seen better or closer. However, we could not tell in advance which haystack the ass would eat from.

We can imagine a thought experiment in which a thousand asses were placed exactly between a thousand sets of haystack pairs. Although any individual ass would remain unpredictable, we would expect that about half would turn to the right and half to the left, or that there was a 50-50 chance of an ass picking either haystack. In statistical terminology, we would say that the chance that an ass will eat from a given haystack is 50 percent. In each decision problem there are one or more statements or claims to be tested. Such a claim is called an *hypothesis*.

Faced with a decision problem, you may take various courses of action. For example, a typical decision problem is when a stock broker recommends that you buy a certain stock. The stock broker either implicitly or explicitly claims that he has a system for picking winners in the stock market. Suppose he claims, in fact, that his system will pick winners most of the time. He argues to your satisfaction that with this high a percentage, you will surely make money. Because the stock broker makes money in commissions, he is trying to interest you in buying a number of stocks according to his system. You are a bit skeptical of his claim and decide to test it on a few stocks which you will follow in the newspaper but will not actually buy. You ask the stockbroker to give you the names of 10 stocks that he recommends for the purpose of seeing how many go up over the next three months. Faced with a real-life decision problem of this kind, you may take either one of two courses of action. You may wait until the end of three months and observe how many of the 10 stocks go up and then make your decision as to the broker's claim by judging these results, or you may plan in advance how many stocks must go up before you will invest your money in such a system. Both of these procedures are used in decision problems. Sometimes it is better to wait and see what happens before making a decision; at other times it is better to plan your behavior beforehand. The science of statistics covers both of these courses of action. In the first case, we use statistical procedures of data analysis; we are concerned with drawing valid conclusions from experimental data or with making proper decisions from observed facts. In the second case we use statistical theory to plan experiments so that the data is gathered with reference to the method of analysis. In any case the proper function of statistics is the making of correct conclusions or decisions from observational data. It is understood that the observational data is influenced to a greater or lesser extent by uncertainty, that is, by probability factors.

Let us now develop a statistical model for making decisions. Because a decision is based on the testing of statements or claims, this part of statistics is called the testing of hypotheses. Let us now go back to our example of a stockbroker who claims he has a method of picking winners, that is, stocks that go up say within three months. How can we test him to see if he has the ability he claims. We want to make our test on the basis of the 10 stocks that he picked as winners. We are skeptical and suspect that he cannot pick a winner. That is, we believe that he will just guess randomly, so that in terms of probability our belief is

$$P(\text{winner}) = P(\text{loser}) = 0.5$$

However, suppose that at the end of three months all of the 10 stocks he selected have gone up, that they are all winners. We would be much less skeptical and wonder how he does it. If we are especially stubborn, we may then ask him to give us a list of 100 stocks, and if they are all winners at the end of three months we would certainly give in and admit he has some system and would invest our money gladly, all the way. Let us analyze why we have made this decision. As we shall see this reasoning is central to decision-making.

The occurrence of 10, or even 100, correct choices by the stockbroker does not prove absolutely that his system really works. He could be guessing randomly, and the fortunate results were a streak of luck. However, on that assumption it would have to be admitted that the stockbroker is extremely lucky. It is very improbable that a random guessor would be right 100 times in a row. Thus if we believe that he is guessing, we must also believe that an extremely unlikely event has occurred. Note that what makes the 100 correct choices improbable is our assumption that the stockbroker is guessing. If on the other hand if the stockbroker had inside information, and picked only stocks that were being manipulated (as some were in the 1920's), we would assume that he would always be correct, and therefore the raw data of 100 correct choices would not be improbable.

In any case, the raw data is 100 correct choices. The results by themselves are neither probable nor improbable. They have a probability only in view of our prior belief about the probability of the stockbroker being correct. For example, if we believe $P(\text{winner}) = 0.5$ then the results are extremely improbable, whereas if we believe that $P(\text{winner}) = 1.0$ then the results are extremely probable. It seems sensible to adopt an assumption (or hypothesis) that makes the data reasonable than one that makes the data bizarre. We can think of our belief that $P(\text{winner}) = 0.50$ as an hypothesis about the stockbroker guessing. On the basis of this hypothesis, we would expect him to be right about 50 out of 100 times. The fact that he was right 100 out of 100 times is extremely different from our prediction is evidence that the initial hypothesis was incorrect, and we would say that his being right 100 out of 100 times was significant.

We call our original, or initial, hypothesis the *null hypothesis* and abbreviate it by H_o. We then collect data. If the data seems probable with respect to H_o, we say the data is *not significant* and accept H_o. On the other hand, if the data seems improbable with respect to H_o, we say that the data is *significant* and reject H_o. But how unlikely must the data be in order to be significant and so reject our null hypothesis? The answer to this question must be based on a judgment made by the investigator. A sufficiently small probability must be one small enough to convince the investigator that the null hypothesis is untenable. Common usage says that a sufficiently small probability is one given by 0.05, or 5 times out of 100, or 1 out of 20. Thus data that occurs 1 time out of 20 when the null hypothesis is true will be regarded as sufficiently improbable so as to reject the null hypothesis. In other words, 1 time out of 20 (on the average) data will be judged as significant even when it is not. The result is a mistake, and is called a *Type I Error.*

The value 0.05 for the probability of rejection is called the *level of significance,* and is denoted by α. It is important to remember that the choice of 0.05 for α is arbitrary. It represents one value that is often used. Other values that are also used are $\alpha = 0.025$ and $\alpha = 0.01$. The particular value of α selected depends upon what one risks by accepting or rejecting the hypothesis.

Type I and Type II Errors

Ancient man distinguished between two types of numbers: peaceful ones like 2, 4, 6, 8 and the warlike ones that were in between. For example, if there were 6 animals and two people having equal claim on them, it was easy to give 3 animals to each and keep peace. If there were 5 animals, 2 could be given to each, but there may have been a fight over the one remaining. Thus the word "even" means flat and smooth, and an even number of anything can be divided into two piles of exactly the same height, that is, of even height. The even number is one that has the property of equal shares. The word "odd" comes from a word meaning "pointed," and an odd number of anything can be divided only if one pile is higher or more pointed than the other. The odd number is one that has the property of "unequal shares." The expression "odds" in betting implies the wagering of unequal amounts of money.

In mathematics it is convenient to say that if two numbers are both even, they are of the *same parity,* and if two numbers are odd, they also are of the *same parity.* However an even number and an odd number, grouped together, are of *different parity.* The usefulness of this convention is based on the following:

(1) The sum of two even numbers is even.
(2) The sum of two odd numbers is even.
(3) The sum of an even and odd number is odd.
(4) The sum of an odd and even number is odd.

Using the symbols E for even and O for odd, these are

(1) $E + E = E$
(2) $O + O = E$
(3) $E + O = O$
(4) $O + E = O$

Instead of four statements, the concept of parity enables us to say the same in two statements:

(1) Same parities add to even.
(2) Different parities add to odd.

When, in any situation, same parities always yield one result and different parities yield the opposite result, we say that parity is conserved.

Let us now apply this convention to *decision theory.* Whenever we consider an hypothesis and a decision, the hypothesis can be either true or false and the decision can be either accept or reject. We pair a hypothesis and a decision. If the hypothesis is true and the decision is acceptance, we say they are of the same parity. Likewise if the hypothesis is false and the decision is rejection, they also are of the same parity. However, a true-hypothesis and reject-decision are of different parity. Likewise a false-hypothesis and a accept-decision are of a different parity. Thus whenever we consider an hypothesis and a decision there are four possible outcomes:

(1) A true hypothesis can be accepted.
(2) A false hypothesis can be rejected.
(3) A true hypothesis can be rejected.
(4) A false hypothesis can be accepted.

Using the symbols T for true, F for false, A for accept, and R for reject, these statements are

(1) $T + A$ = No error
(2) $F + R$ = No error
(3) $T + R$ = Error (Type I)
(4) $F + A$ = Error (Type II)

Instead of these four statements, we can say

(1) Same parities result in no error.
(2) Different parities result in error.

As we have stated previously a Type I Error is the case in which the hypothesis is true but we reject it. A Type II Error, then is the case in which the hypothesis is false but we accept it. Notice that we can make at most one of these two errors in any decision.

Exercises

1. A null hypothesis is rejected if
    (a) it makes the probability of the data adequately high
    (b) it makes the probability of the data adequately low
    (c) the data is insignificant
    (d) no data is available

2. Assuming the null hypothesis is true the probability of rejecting the hypothesis is
    (a) $50 - 50$
    (b) α
    (c) $1 - \alpha$
    (d) 0

3. Assuming the null hypothesis is true, significant data (at the 5 percent level) is
    (a) unlikely
    (b) likely
    (c) 95 percent likely
    (d) 100 percent likely

4. If we reject the null hypothesis, we cannot make a
    (a) correct decision
    (b) incorrect decision
    (c) Type I Error
    (d) Type II Error

5. On the basis of an experiment, it was decided that a drug claimed to be harmless produced significant side effects. A Type I Error was committed if
    (a) Later definitive experiments showed the drug was extremely bad.
    (b) Later definitive experiments showed the drug was harmless.

6. Data is called significant if, with respect to the claim ($=$ null hypothesis), the data is
    (a) very likely
    (b) very unlikely
    (c) neither likely nor unlikely
    (d) made up of large numbers

7. After each of these situations write either Type I Error, Type II Error, or no error.
    (a) Accepted a false claim
    (b) Rejected a true statement
    (c) Accepted a true hypothesis
    (d) Rejected a false statement

CHAPTER 12

BAYESIAN DECISION MAKING

Bayes rule for flipping trees

Suppose you have three sets of cousins all of whom live in the same neighborhood: the Smiths with two girls, the Jones's with a girl and a boy, and the Browns with two boys, which make up six cousins altogether. We assume that a person has an equal chance of coming in contact with any one of the three families, and once contact is made there is an equal chance of meeting either member of the family. These probability assumptions are summarized in the tree diagram.

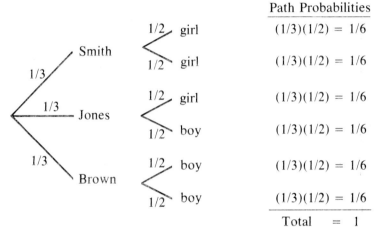

On the right of this tree diagram we have computed the path probabilities, which sum to one. We now want to flip the tree diagram to obtain:

101

A path probability in the flipped tree can be obtained by finding the corresponding path probability in the original tree. For example, the path Jones-boy in the original tree has the same probability as the path boy-Jones in the flipped tree. In this simple example, all path probabilities are the same, namely 1/6.

Next we must compute the probabilities in the branches of the flipped tree. We must give the girl branch a probability equal to the sum of the probabilities of all the paths that include the girl branch; that is, we assign probability (a) as 1/6 + 1/6 + 1/6 = 1/2. Similarly, the assignment (b) is 1/6 + 1/6 + 1/6 = 1/2. Having assigned (a) and (b) we can now get assignments (c), (d), (e), (f), (g), (h). Since we know the path probability of girl and Smith is 1/6, which is the product (a)(c), and since we know (a) is 1/2, it then follows that

$$1/6 = 1/2 \text{ (c)}$$

so (c) = 1/3. Similarly, (d) = 1/3, (e) = 1/3, (f) = 1/3, (g) = 1/3, (h) = 1/3. Thus the flipped tree is

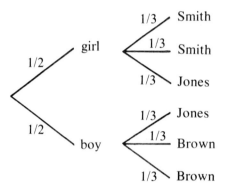

Suppose now that somebody meets one of your cousins at a party and the cousin is a girl. If he didn't catch her last name, what is the probability that it is Smith? From the flipped tree we see that from girl there are two branches for Smith, each with probability 1/3, so the required probability is 1/3 + 1/3 = 2/3. The method we have used to find this probability is called *Bayes Rule* or *Bayes Theorem*.

Let us give another example. Suppose you are interested in all families with two children. As a matter of convenience we call a family with two girls a Smith family, a family with one of each sex a Jones family, and a family with two boys a Brown family. Suppose we meet a girl from a two-child family. From our above analysis, we have found that the probability is 2/3 that she comes from a Smith family, or in other words the probability is 2/3 that her sibling is also a girl. But we know that births are independent, and the probability of a girl is (approximately) 1/2, so the correct answer is that the probability is 1/2 that her sibling is a girl. Why has Bayes theorem given us the result 2/3, when we know the correct result is 1/2?

The answer is that Bayes theorem is correct, but our application of it to the case of two-child families is wrong. We know that the events (where G = girl, B = boy)

GG, GB, BG, BB

are (approximately) equally likely, so a Jones-type family which is made up of either *GB* or *BG* is twice as likely as either a Smith-type family (*GG*) or a Brown-type family (*BB*). Thus our original tree is

Path Probabilities

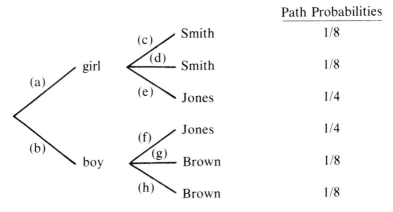

and the flipped tree is

Path Probabilities

We have

$$(a) = 1/8 + 1/8 + 1/4 = 1/2$$

$$(a)(c) = 1/8 \quad \text{so} \quad (c) = \frac{1/8}{1/2} = 1/4$$

$$(a)(d) = 1/8 \quad \text{so} \quad (d) = \frac{1/8}{1/2} = 1/4$$

Hence if we meet a girl from a two-child family, the probability is

$$(c) + (d) = 1/4 + 1/4 = 1/2$$

that she is from a Smith-type family, or in other words the probability is

1/2 that her sibling is also a girl. This represents the correct application of Bayes theorem to the problem of two-children families.

Reaction to experimental evidence

Many psychologists have investigated the intuitive reactions of subjects to experimental evidence of a probabilistic nature. Let us pose the following problem:

We have two bags each with five marbles. The first bag contains three blue marbles and two white marbles, and we refer to this bag as the predominantly blue bag. The second bag contains only one blue marble and four white marbles, and we refer to this bag as the predominantly white bag. The bags are identical in appearance. One bag is drawn at random, and then one marble is drawn at random from that bag. The marble is blue. What is the probability that it came from the predominantly blue bag?

Various answers are given, but after a while a consensus emerges to the effect that the evidence is meager so that the probability will only be slightly better than the probability of 1/2.

We can apply Bayes rule as follows. We construct the tree diagram:

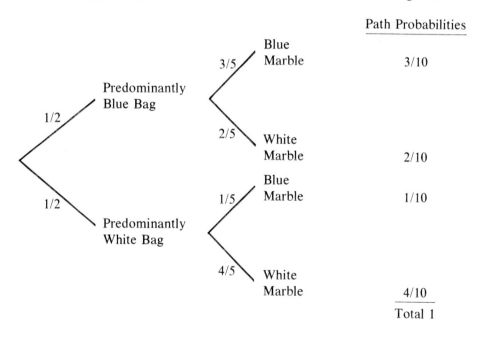

The flipped tree diagram is

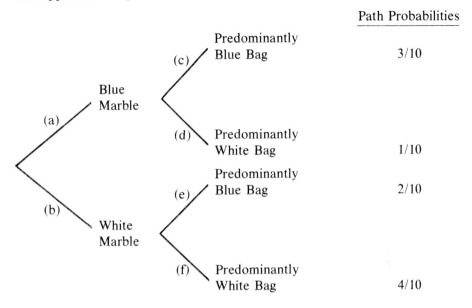

We have

$$(a) = 3/10 + 1/10 = 4/10$$

$$(a)(c) = 3/10 \quad \text{hence} \quad (c) = \frac{3/10}{4/10} = \frac{3}{4} = 0.75$$

Thus given the evidence of a blue marble, the probability of the bag being the predominantly blue bag is 0.75. Thus the evidence has increased the probability from 0.50 (namely, the probability of drawing the predominantly blue bag) to 0.75 (namely, the probability of drawing the predominantly blue bag given the marble drawn is blue).

Instead of constructing the flipped tree diagram we can easily compute the required probability from the original tree diagram as follows:

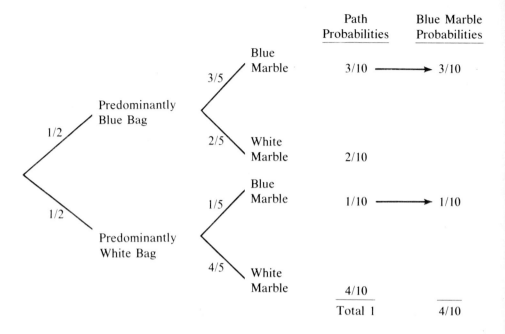

The required probability is found by dividing the probability 3/10 by the column total 4/10, that is, the probability of a blue bag given a blue marble is drawn is

$$\frac{3/10}{4/10} = \frac{3}{4} = 0.75$$

Hypothesis testing

A decision problem in which one of two actions must be chosen is called a problem in *hypothesis testing*. A trial in court is a problem of this type, and the two actions are acquit or convict. Another problem in hypothesis testing is the inspection sampling of manufactured goods, where the two actions are accept or reject the lot. In such cases, one action is appropriate to certain possible states of nature, and the other action is appropriate to other possible state of nature.

The simplest kind of problem in hypothesis testing is one in which there are only two possible states of nature. For example, a child may either be well or sick; these represent the two states of nature. The two possible actions are either not call a doctor or call a doctor. A mother at home with her child must make a decision whether to call a doctor or not. Clearly the action "not call a doctor" is appropriate for the state "well" while "call a doctor" is appropriate for the state "sick." Should the mother call the doctor or not? To answer this question we need to know the relative cost of a wrong decision and the probabilities of the child being either well or sick. Suppose from long experience the mother knows that each time her child wakes up in the morning and says he is sick, he is actually sick only one time in four. Her prior probability of her child being sick is then 0.25 and being well is 0.75. The mother's null hypothesis is that the child is well.

If she calls the doctor when the child is really well, she is making a Type I Error. Let us assign a relative cost of 1 unit to making this Type I Error; this cost would be the doctor's fee for a service rendered that was not needed plus the psychic cost of the embarrassment of needlessly taking up the doctor's time when he is needed on other cases. This relative cost of 1 unit is called the regret r of the Type I Error.

If she doesn't call the doctor when the child is really sick, she is making a Type II Error. Let us assign a relative cost of 10 units to making this Type II Error; this cost would be the extra fees required for not catching the sickness in time plus the psychic cost of not properly treating her sick child. This relative cost of 10 units is called the regret R of the Type II Error.

These probabilities and regrets are shown in the table:

STATE ACTION	Well $P(\text{Well}) = 0.75$	Sick $P(\text{Sick}) = 0.25$
Don't call doctor	0	$R = 10$
Call doctor	$r = 1$	0

We calculate the expected loss for each action by summing the losses in each row weighted by their appropriate probabilities:

Expected loss for not calling doctor = $0(0.75) + 10(0.25) = 2.50$
Expected loss for calling doctor = $1(0.75) + 0(0.25) = 0.75$

We see that the expected loss is less in the case of the action of calling a doctor, so the optimal decision is that action.

Up to now we have assumed that the mother has taken no statistical information as to her child's condition. The mother has available only a thermometer which she has trouble reading so there is some question as to the accuracy of her measurement. She takes her child's temperature and finds that it is 99.5° as far as she can tell. Let us suppose that temperature is normally distributed, with a mean of 99° if the child is well and with a mean of 101° if the child is sick, and with the same standard deviation equal to 1° in either case. What now is the best action, given the temperature measurement of 99.5°? At first, we might reason as follows. We draw each of the two distributions as shown in the diagram:

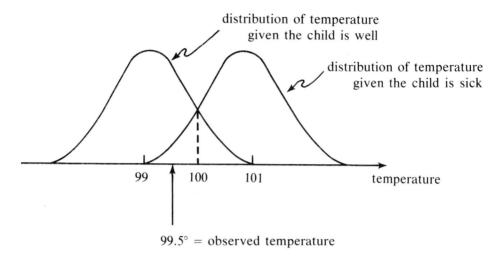

99.5° = observed temperature

Using the most simple type of reasoning, we would accept the hypothesis which has the greater probability of generating the observed temperature of 99.5°. At the point 99.5° on the temperature axis, it is seen that the curve for the case of the child being well lies above the curve for the sick case, and thus on this criterion we would make the decision that the child is well and not call the doctor. In the above diagram the two curves intersect at a point corresponding to a temperature of 100°. To the left of this point the well curve lies above the sick curve, whereas to the right of this point the sick curve lies above the well curve. Hence 100° represents the decision point, and so we can formulate the following decision rule: The mother should not call the doctor if the temperature is below 100° and should call the doctor if the temperature is above 100°. The 100° mark is the break-even point, so either of the two actions would be OK at this point.

Decision point

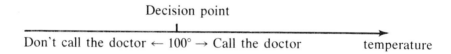

Don't call the doctor ← 100° → Call the doctor   temperature

All of a sudden, we say "Wait a minute!". In our analysis we have given exactly equal weighting to the two normal curves (the cases of well and sick), but we really know that 75 percent of the time the well one applies as to only 25 percent of the time for the sick one and we also know that our regret is $R = 10$ if we don't call when the child is sick versus $r = 1$ if we call the doctor when the child is well. The fact that the child is well 75 percent of the time argues for pushing the decision point to the right; on the other hand the fact that our regret is 10 times as big for not calling the doctor when the child is sick argues for pushing the decision point to the left.

The naive decision point is the average of the mean temperature of $\mu_0 = 99°$ when the child is well and the mean temperature $\mu_1 = 101°$ when the child is sick:

$$\text{Naive decision point} = \frac{\mu_0 + \mu_1}{2} = \frac{99 + 101}{2} = 100°.$$

For normal distributions with $\mu_0 < \mu_1$ and with a common σ, it turns out that the decision point should be moved by an amount

$$\frac{\sigma^2}{\mu_1 - \mu_0} \ln \frac{P(\text{null hypothesis true})}{P(\text{null hypothesis false})}$$

in order to take into consideration the prior probabilities, and the decision point should be moved by an amount

$$\frac{\sigma^2}{\mu_1 - \mu_0} \ln \frac{\text{Regret of Type I Error}}{\text{Regret of Type II Error}}$$

in order to take into consideration the regrets. In our case the decision point should be moved by an amount

$$\frac{1}{101 - 99} \ln \frac{P(\text{well})}{P(\text{sick})} = (1/2)\ln \left(\frac{0.75}{0.25}\right) = \left(\frac{1}{2}\right) \ln 3$$

due to the fact that the child is well 3 times as often as he is sick, and the decision point should be moved by an amount

$$\frac{1}{101 - 99} \ln \frac{r}{R} = \left(\frac{1}{2}\right) \ln \frac{1}{10} = -\frac{1}{2} \ln 10$$

due to the fact that our regret for a Type II Error is 10 times our regret for a Type I Error. Here "ln" stands for the natural logarithm. A short table of natural logarithms is:

n:	1.0	1.5	2.0	2.5	3.0	4	5	6	7	8	9	10
$\ln n$:	0.0	0.4	0.7	0.9	1.1	1.4	1.6	1.8	1.9	2.1	2.2	2.3

We must move the decision point

$$\left(\frac{1}{2}\right) \ln 3 = \left(\frac{1}{2}\right) (1.1) = 0.55$$

due to the prior probabilities and also move the decision point

$$-\left(\frac{1}{2}\right) \ln 10 = -\left(\frac{1}{2}\right) (2.3) = -1.15$$

due to the regrets. That is, we must move the decision point 0.55° to the right, and 1.15° to the left, which makes a net move of $0.55 - 1.15 = -.60°$, that is, a move of 0.60° to the left. Thus the final decision point is

$$\text{Decision Point} = \text{Naive Decision Point} + \text{Net Adjustment} = 100° - 0.60°$$
$$= 99.4°.$$

The decision rule is

$$99.4°$$
Decision Point

———————————|——————————————————————→

Don't call the doctor ←—|—→ Call the doctor   temperature

Since the mother observed a temperature of 99.5° she should call the doctor.
    In summary, suppose we have

Null hypothesis: Normal with mean μ_0 and standard deviation σ
Alternate hypothesis: Normal with mean μ_1 and standard deviation σ where

$$\mu_1 > \mu_0$$

Prior probability of null hypothesis being true $= p$
Prior probability of null hypothesis being false $= q = 1 - p$

Regret of Type I Error $= r$
Regret of Type II Error $= R$

Then the decision point is given by

$X =$ naive decision point $+$ correction for prior probabilities and for regrets

$$= \frac{\mu_0 + \mu_1}{2} + \frac{\sigma^2}{\mu_1 - \mu_0} \ln \frac{pr}{qR}$$

and the decision rule is

Decision
Point

———————————————————|————————————————————→

Accept null hypothesis X Reject null hypothesis   temperature
if observation falls on   if observation falls on
this side of decision point.   this side of decision point.

    Let us recompute the decision point for our example. We have

$$X = \frac{99 + 101}{2} + \frac{1}{101 - 99} \ln \frac{0.75\,(1)}{0.25\,(10)} = 100 + \left(\frac{1}{2}\right) \ln \frac{3}{10}$$

$$= 100 + \left(\frac{1}{2}\right) \ln 3 - \left(\frac{1}{2}\right) \ln 10 = 100 + \left(\frac{1}{2}\right)(1.1) - \left(\frac{1}{2}\right)(2.3) = 99.4$$

The decision point may be interpreted as follows. We multiply the normal curve in
the case the child is well by

$$P(\text{well})(\text{Regret for Type I Error}) = pr = (0.75)(1) = 0.75$$

and we multiply the normal curve in the case the child is sick by

$$P(\text{sick})(\text{Regret for Type II Error}) = (1 - p)R = (0.25)(10) = 2.5.$$

Then the decision point is that value of temperature corresponding to the point of intersection of the two new curves, as seen in the diagram:

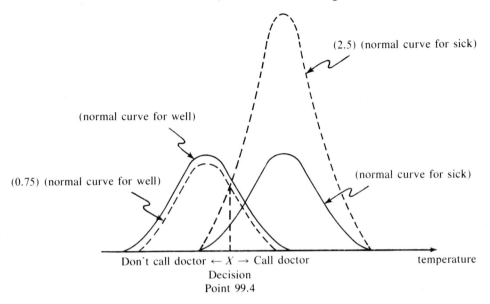

(2.5) (normal curve for sick)

(normal curve for well)

(0.75) (normal curve for well)

(normal curve for sick)

Don't call doctor ← X → Call doctor   temperature
Decision
Point 99.4

The probability of a Type I Error is the area under the normal curve for well to the right of the decision point X. This probability is

$$\alpha = P\left(z > \frac{X - \mu_0}{\sigma}\right) = P(z > 0.4) = 0.34$$

The probability of a Type II Error is the area under the normal curve for sick to the left of the decision point X. This probability is

$$\beta = P\left(z < \frac{X - \mu_1}{\sigma}\right) = P(z < -1.6) = P(z > 1.6) = 0.05$$

These areas are shown in the diagram:

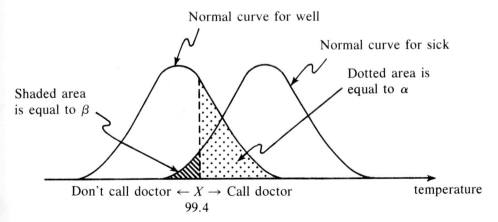

Normal curve for well

Normal curve for sick

Dotted area is
equal to α

Shaded area
is equal to β

Don't call doctor ← X → Call doctor   temperature
99.4

The tree diagram is:

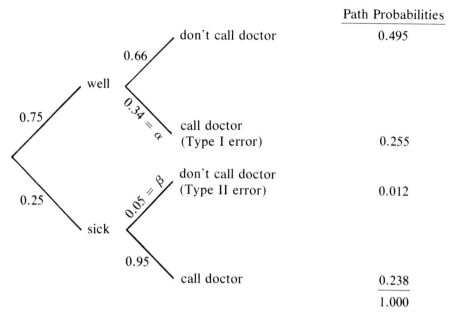

The flipped tree diagram is:

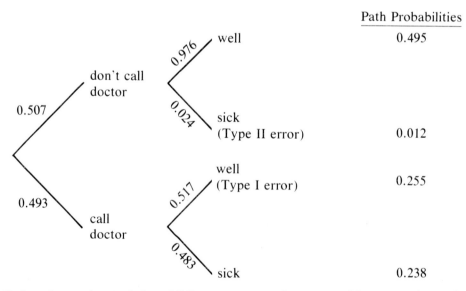

Before the mother took her child's temperature there was a 25 percent chance he was sick (i.e. the prior probability $P(\text{sick}) = 0.25$) as seen on the original tree diagram. After the mother took her child's temperature and found that the observed temperature of 99.5° fell in the critical region to the right of the decision point of 99.4°, her action would be to call the doctor. From the flipped tree diagram we see that probability that the child is sick given that this action is taken is now 0.483. In other words the posterior probability of sickness, namely 0.483, is almost twice as great as the prior probability of sickness, namely 0.250. That is the observed temperature of 99.5° was responsible for revising the probability of sickness from 0.25 up to 0.483.

Finally let us compute the expected loss. The path probability for a Type I Error is 0.255 and the regret for a Type I Error is 1. The path probability for a Type II Error is 0.012 and the regret for a Type II Error is 10. Hence the expected loss is

$$(0.255)(1) + (0.012)(10) = 0.375$$

We recall that the smallest expected loss that could be incurred without taking the temperature was 0.75. By taking the temperature we see that the expected loss is reduced by one-half.

Exercises

1. The probability of getting tails with a fair coin is 0.5. The probability of getting tails with a particular unfair coin is 0.2. The prior probability of the coin being fair is 0.8. The path probability of getting a fair coin and then getting tails is
      (a) 0.40 (b) 0.16 (c) 0.04 (d) 0.50

2. Flip the tree

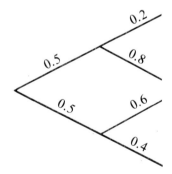

3. In a course, students were either silent S or responsive R. Seventy percent of the S students got an A while 30 percent of the R students got an A. Fill in as much as you can on the tree diagram.

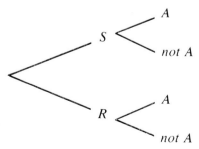

4. If, in the above problem 45% were *S* and 55% were *R*, then what is the probability that a silent student gets an *A*?

5. In the above problem, which is the probability that an *A* student is silent? (First draw the flipped tree.)

6. If the regret for a Type II Error is much greater than for a Type I Error, then we should favor calling the doctor over not calling him. TRUE or FALSE? [Here assume that the null hypothesis is well.] First identify the Type I and Type II Errors (i.e. which is "calling when well" and "not calling when sick".)

7. Suppose it rains 40 percent of the days and shines 60 percent of the days. On rainy days the barometer erroneously predicts shine 20 percent of the time, and on shiny days the barometer erroneously predicts rain 30 percent of the time. After looking at the barometer and seeing it predict rain, what is the posterior probability of rain?

(a) 0.40   (b) 0.36   (c) 0.84   (d) 0.64

8. An archaeologist has to classify skulls as Tribe *A* or Tribe *B* on the basis of their width. The populations of skull widths are normally distributed as follows:

	Mean	Standard Deviation
Tribe *A*	12 cm.	2 cm.
Tribe *B*	15 cm.	2 cm.

At first the archaeologist computed $\dfrac{12 + 15}{2} = 13.5$, and classified a skill as Tribe *A* if its width is less than 13.5 and as Tribe *B* if its width is greater than 13.5. Then he found that Tribe *A* was a much more numerous tribe than Tribe *B* and so he should move his decision point from 13.5 to (pick best answer):

(a) 14.5   (b) 12.5   (c) 13.5 (i.e. the same)   (d) 12.0

9. A miser has great regrets if he calls the doctor when he is well but small regrets if he doesn't call the doctor when he is sick. An ordinary person might call the doctor only if his temperature is more than 99.5° but the miser would call only if his temperature was more than (pick best answer):

(a) 98°   (b) 98.6°   (c) 99.5°   (d) 101°

CHAPTER 13

ANALYSIS OF DECISIONS UNDER UNCERTAINTY

Conditional analysis and the payoff table

The relevant consequences of any business decision, whether monetary or nonmonetary, always lie in the future; nothing we can do today will change the past. An essential part of the analysis of any decision problem, therefore, is a forecast. Difficult as it is to get accurate historical data, forecasts are even more likely to be in error. It is a rare occasion when events materialize exactly as planned. One way to deal with the problem of uncertain forecasts is to make an analysis of a decision problem on the basis of several alternative estimates of the unknown events. If, for example, we are considering the introduction of a new product which will entail an increase in fixed costs, we might calculate the profitability of the new product for several sales volumes. Sometimes such an analysis points to a single, definite decision. It might turn out, say, that even if the proposed new product had the highest sales volume considered of all possible, its contribution would still not cover the increment in fixed costs. In this case uncertainty presents no difficulties; while the "true" sales volume is not known, it is known that, no matter what this true volume is, introduction of the product would be an unprofitable decision. More commonly, it will turn out that one act would be preferable if some events occur, while some other act or acts would be better if other events occur. The decision-maker, not knowing what event will in fact occur, must somehow balance one act off against another to reach an un-equivocal decision.

In this chapter, we will examine the two major questions raised by the presence of uncertainty in a decision problem:

1. What is a reasonable criterion for choosing among acts when their consequences cannot be predicted with certainty?
2. Rather than making an immediate choice, should additional information be obtained in order to improve the forecasts of consequences?

This chapter will be concerned with the selection of a criterion.

A retailer wishes to decide how many units of a particular commodity to order. Since the commodity is perishable and cannot be kept in stock for more than a day, the retailer does not want to order more than a day's supply. On the other hand, since each unit costs the retailer only $1 while he sells it for $5, each demand he fails to satisfy as a result of underordering represents a loss to him of $4 in gross margin, to say nothing of possible loss in customer good will. If the retailer knew for certain exactly what his demand would be on a particular day, it is clear that he would want to order exactly enough units to meet the demand, no more and no less. Unfortunately, the world being what it is, the retailer does not know what demand will actually be. And yet he must decide on a particular number to stock.

In order to simplify the arithmetic in this example, we will hereafter assume that the retailer's shelf space limits him to a stock of 5 units, so that in any event he will not order more than that number. Thus his possible decisions are to stock 0, 1, 2, 3, 4, or 5 units. One possible way for the retailer to proceed is to analyze his problem under several alternative assumptions as to what his demand will be. We will call such an approach a conditional analysis because of its "if, then" character; it does not try to predict what the outcome of a particular act will in fact be, but only what it will be *conditional* on a particular event occurring. As an example, let us take the event "demand for three units" and determine what the retailer's gross profit would be from each possible act, or stock level. First of all, we note that the retailer's total cost will be $1 per unit times the number of units ordered. His revenues will be $5 × 3 = $15 providing he stocks at least three units; otherwise it will be $5 times the number of units stocked, with the remaining demands being unsatisfied. We can summarize this situation in the following table:

TABLE 1
Gross Profit if Demand is for Three Units

	Act (Stock Level)					
	0	1	2	3	4	5
Revenue	$0	$5	$10	$15	$15	$15
Cost	0	1	2	3	4	5
Gross Profit	$0	$4	$ 8	$12	$11	$10

There is no real difference between the conditional analysis of a decision problem under uncertainty and the usual economic analysis of a decision problem "under certainty." Only cost and revenues which actually depend upon what act is selected should be considered in the analysis. This means that we look only at future costs and revenues, ignoring "sunk" costs; and that even the future costs or revenues which would be the same regardless of the act selected need not be considered. In our retailer example, for instance, we can ignore any costs and revenues other than those listed in Table 1 if we can validly make the assumption that all other costs and revenues would be unaffected by the retailer's choice of a stock level. We will in fact make this assumption in our further discussions of this example, although it is unlikely to be strictly correct. The retailer's failure to meet a demand through understocking may adversely affect his sales of other items or his future sales of all items, and hence his true loss in revenue from an unsatisfied demand would be greater than $5. Again, as in the case of an economic analysis under certainty we have included in Table 1 only those consequences which have been measured in monetary terms. While it is often possible to express noneconomic factors in terms of their monetary equivalents, this is not always practical. When important noneconomic considerations have been excluded from the formal analysis, they must be considered judgmentally in the interpretation of the analysis.

Continuing as in Table 1, we can calculate the conditional gross profit of each feasible stock level for various other levels of demand. The results of such calculations are summarized in Table 2.

TABLE 2
Payoff Table for Retailer Example

Demand	Act (Stock Level)					
	0	1	2	3	4	5
0	$0	−$1	−$2	−$3	−$4	−$5
1	0	+ 4	+ 3	+ 2	+ 1	0
2	0	+ 4	+ 8	+ 7	+ 6	+ 5
3	0	+ 4	+ 8	+12	+11	+10
4	0	+ 4	+ 8	+12	+16	+15
5 or more	0	+ 4	+ 8	+12	+16	+20

A table such as Table 2, in which the consequences of a number of acts are evaluated for a number of events, is called a *payoff table*. Sometimes we will instead refer to the table in terms of the criterion used to evaluate the consequences; for example, we could call Table 2 a *gross profit table*. Aside from the calculation of conditional payoffs, there are two main questions involved in the preparation of a payoff table: (1) What acts should be included? (2) What events should be included?

Acts in Payoff Tables. Quite often, as in our retailer example, it will be relatively easy to list all of the feasible alternatives in a decision problem. In other problems, however, it will not be so easy to recognize what the alternatives are. No rule can be given for determining what alternatives should be considered in a particular problem. This determination must be a matter of imagination and judgment in the particular case. The best any analysis can do is to identify the best act among those examined; there might be still better acts which have not been included in the analysis.

Events in Payoff Tables. Somewhat more specific advice can be given on the events to be included in a payoff table. First of all, it should be clear from the retailer's example that the choice of events must make it possible to state in an unambiguous way the conditional payoffs of any act given each event. For example, it would not do to have "demand for 3 or 4 units" as an event in the retailer example, since it would not be possible to state unambiguously what the gross profit would be in this event if either 4 or 5 units were stocked. On the other hand, we can lump together all demands for five or more units, since the limitation on the retailer's shelf space insures that the payoff for each feasible act will be the same for all demands greater than or equal to five.

The second requirement for a listing of events is similar to the first: it must be logically impossible for two or more of the events contained in a given listing to occur simultaneously. For example, it would not do to include both "demand for 5 units" and "demand for 5 or more units" in a listing of events, since if demand is for 5 units, this occurrence cannot be assigned unambiguously to one or the other of the two events. The reason for this second requirement is to avoid the possibility of "double counting" of some of the payoffs by virtue of their being listed twice. Another way of stating that two events cannot occur simultaneously is to say that they must be *mutually exclusive*.

The third requirement for a listing of events is that it must be complete, in that some one of the events included on the list must occur. This requirement is sometimes stated by saying that the listing must be *collectively exhaustive*. This requirement is included to insure that all consequences of an act which might occur are brought into proper consideration. If, in the retailer example, we substituted "demand for 5 units" for "demand for 5 or more units," the listing would not be collectively exhaustive unless the retailer were absolutely convinced that demand would not exceed 5 units, since the payoff table would not show the consequences of demand for 6 or more units. (The fact that these consequences are the same as if demand was for 5 units does not contradict the last sentence, since it is only if demand for 6 or more units is considered specifically that we can be sure that the consequences are the same.)

If the number of acts and/or the number of events is relatively small, the payoff table is an exceptionally convenient way of organizing the economic data bearing on a decision problem under uncertainty. The student may well wonder, however, what we would do if the retailer wanted to consider any stock level up to, say, 5,000 units, and believed that his potential demand was subject to a similarly substantial range. Clearly in such a case the preparation of a payoff table would be impractical.

In a "big" problem in which the payoff table could not be constructed, one strategy would be to try to summarize the economic data in the form of algebraic equations. In the retailer's problem, for example, we could use the following equations:

$$\text{Gross Profit} = \begin{cases} 5D - S & \text{if } D \le S \\ 4S & \text{if } D > S \end{cases}$$

where D represents the quantity demanded (event) and S represents the quantity stocked (act). These equations express exactly the same information as Table 2, since they enable us to find the gross profit for any act-event combination. However, by expressing this information algebraically, the equations are much more compact and moreover permit the determination of gross profit for many more act-event combinations than would be practical in a payoff table.

Despite the practical advantages of the algebraic form, the payoff table is still important as a conceptual tool. It is often helpful, in beginning the analysis of a decision problem under uncertainty, to imagine what the payoff table would look like, even if it is not practical to list all the acts and events. And in this note, since our objective is to bring about a conceptual understanding of decision-making under uncertainty rather than to teach techniques for "solving" particular problems, we will emphasize simple examples in which the payoff table can actually be constructed in full.

Possible criteria for selecting an act

Although the payoff table is a convenient way to organize the economic data bearing on a decision problem under uncertainty, it presents *conditional information only. We must still find a way of deciding unconditionally on a "best" act.*

One possibility is to concentrate attention on the "most likely" event. If, for example, the retailer in our example felt that demand was more likely to be for 3 units than for any other number, he might look only at the row "demand for 3" and decide to order 3 units. Concentrating on the most likely event is probably a very common way of deciding decision problems under uncertainty in practice.

Another possibility is the "conservative" approach of looking at the worst that can happen for each possible act and then choosing the act for which the "worst" outcome is the most desirable. In the retailer's problem, the most unfavorable event for any act would be "demand for 0 units." Since this event will always result in an out-of-pocket loss so long as the retailer orders any units at all, the "conservative" solution would be to order no units. Any businessman so averse to risk-taking as to favor this approach has no business being in business!

The principal drawback in each of the suggested approaches is that it concentrates on a single consequence of each act to the exclusion of all others. A fully reasoned solution of a decision problem under uncertainty ought to take into consideration all the consequences which might conceivably follow from each act. At the same time, it seems reasonable that not all of the consequences should be treated as equally important. If the retailer in our example considered the event "demand for 5 or more units" as quite unlikely, as well as considering "demand for 3 units" as most likely, then he would want to give more weight to the consequences of a demand for 3 units than to those of a demand for 5 or more units.

In the above discussion we used such phrases as "more likely" and "most likely" in referring to events. Instead of referring to likelihood in qualitative terms of "more" and "less," however, we might put likelihood on a quantitative basis by assigning numbers to events which represent their likelihood. As we know, such numbers are called probabilities.

Leaving until later the question of how probabilities might be assigned to events, let us assume that the retailer in the example we have been discussing has assigned probabilities to potential demands as given in Table 3.

The expected monetary value criterion

Several paragraphs back we indicated that a good criterion for evaluating acts with uncertain consequences ought to reflect: (1) all of the possible consequences of the act; and (2) the relative likelihood of these consequences. One criterion

TABLE 3
Probability Distribution of Demand

Demand	Probability
0	.05
1	.15
2	.30
3	.25
4	.15
5 or more	.10
	1.00

which meets the standards just given is called expected monetary value. The *expected monetary value* of an act is a weighted average of the conditional monetary consequences of that act, using the probabilities assigned to events as the weights. Taking the conditional gross profits from Table 2, and the probabilities from Table 3, we can calculate the expected monetary value of the act "stock 3 units" as in Table 4.

TABLE 4
Calculation of Expected Monetary Value with Stock of 3

Demand	Probability	Conditional Profit	Weighted Profit
0	.05	−$ 3	−$ .15
1	.15	+ 2	+ .30
2	.30	+ 7	+ 2.10
3	.25	+ 12	+ 3.00
4	.15	+ 12	+ 1.80
5 or more	.10	+ 12	+ 1.20
	1.00		+ 8.25

The total of the last column of Table 4, $8.25, is the expected monetary value of the act stock 3 units. We will also call this the *expected profit*. In other problems, we might be dealing with conditional and expected costs.

In the same manner, we can calculate the expected profits of the other stock levels open to the retailer. The results are presented in Table 5. The reader may verify one or more of these figures to insure that the method of calculation is understood.

The act "stock 4 units" has a higher expected profit, $8.50, than any of the other possible acts. If the retailer wishes to use the expected monetary value criterion, therefore, he would decide to stock 4 units.

While the expected monetary value criterion does consider all monetary consequences of a given act and weight them according to their relative likelihoods, there are still some situations where it may not apply. Suppose, for example, that you were offered your choice of two acts. Act 1 is absolutely certain

TABLE 5
Expected Profits of all Possible Stock Levels

Stock Level	Expected Profit
0	$ 0
1	3.75
2	6.75
3	8.25
4	8.50
5	8.00

to result in neither a profit nor a loss. Act 2, on the other hand, will lead either to a profit of $55,000 or an out-of-pocket loss of $45,000, either with a probability of one-half.

Quite possibly you would decide to pick Act 1, even though Act 2 has an expected monetary value of $5,000 versus only $0 for Act 1. There is nothing inherently "irrational" about such a choice, since there is an important difference in riskiness between the two acts which is not reflected in their expected monetary values. It may be desirable to give up $5,000 in expected value to protect against the possible loss of $45,000. Suppose, on the other hand, that you were offered a choice between Act 1 and Act 3, where Act 1 is the same as above ("do nothing") whereas Act 3 leads to either a profit of $0.55 or an out-of-pocket loss $0.45 each with probability one-half. In this case you might well be willing to act in accordance with expected monetary value and choose Act 3, since the difference in riskiness between the two acts is so slight in absolute terms.

Thus, the applicability of the expected monetary value criterion to a particular decision problem will depend upon the relative riskiness of the acts being considered. No hard and fast rule can be given, since matters will largely depend upon the circumstances of the particular decision-maker. A company with substantial financial resources, for example, will be able to assume more risk than one struggling to meet its bills; hence it may be willing to use expected monetary value in a larger number of cases. Later in this chapter we will discuss a criterion which does take into account differences in riskiness between acts: expected utility. The discussion of expected utility will also indicate a test which can always be made to ascertain whether expected monetary value is appropriate in a given decision problem.

Assessment of probabilities

Two main sets of ingredients go into the calculation of the expected monetary value of an act: (1) the conditional monetary values of the act for each possible event; and (2) the probabilities of the events. Determination of conditional values was discussed previously; we will now be concerned with the determination of probabilities. If probabilities are to make sense when used as weights in calculating expected monetary values, they must obey certain fundamental rules. Three rules which must be obeyed by any set of probabilities are discussed in this section. In following sections we will discuss how these rules will help in assigning probabilities in particular problems.

In some event in a given decision problem is considered impossible, then the decision-maker will not want the conditional values of this event to receive any weight at all in an expected value calculation. This can be accomplished by assigning a weight (probability) of 0 to that event. If the event is considered at all possible, however, then the weight should be greater than 0. Hence we arrive at

Rule 1: A probability is a number greater than or equal to 0 assigned to an event.

This rule was used in the assignment of probabilities leading up to Table 4.

It will also be observed that the probabilities in that Table add up to 1.00. There is no real logical reason why these probabilities must add up to 1.00, however; for instance, if we multiplied each probability in that Table by 3, the calculations would appear as in Table 6. Notice that, as the last step in the calculations, we must divide the weighted total of conditional profits by the sum of the weights (3.00) in order to get the weighted average of $8.25.

TABLE 6

Alternate Calculation of Expected Monetary Value with Stock of 3

Demand	Weight	Conditional Profit	Weighted Profit
0	.15	−$ 3	−$ .45
1	.45	+ 2	+ .90
2	.90	+ 7	+ 6.30
3	.75	+ 12	+ 9.00
4	.45	+ 12	+ 5.40
5 or more	.30	+ 12	+ 3.60
	3.00		+$24.75

$$\text{Weighted Average} = \frac{\$24.75}{3.00} = \$8.25$$

By making the sum of the weights equal to 1.00, we can eliminate this last step in the calculations, since with weights adding up to 1.00 the weighted total and the weighted average of a set of numbers are the same. For this reason we will adopt

Rule 2. The sum of the probabilities assigned to a set of mutually exclusive and collectively exhaustive events shall be 1.

It will be noticed in Table 2 that the conditional gross profit of the act stock 3 units is the same for each of the three events demand for 3 units, demand for 4 units, and demand for 5 or more units. This being the case, it is possible to group these three events into a single event demand for 3 or more units without losing any relevant information about the act stock 3 units. This is done in Table 7.

TABLE 7
Alternative List of Events for Act 3 Units

Demand	Conditional Profit
0	−$ 3
1	+ 2
2	+ 7
3 or more	+ 12

If now we were to calculate the expected monetary value of stocking 3 units using Table 7 rather than Table 2, what probabilities should be assigned to the events? It should be evident that the events that demand is for 0, 1, or 2 units are identical with those of Table 2 and hence should get the same probabilities, i.e., those given in Table 3. Moreover, if we are to satisfy the requirements just mentioned as well as Rule 2, the weight assigned to the event demand for 3 or more units must equal the sum of the weights of the three events of which it is composed. This requirement can be generalized to

Rule 3. The probability of an event which is composed of a group of mutually exclusive events is the sum of their probabilities.

The words "mutually exclusive" in Rule 3 are necessary to avoid double-counting of some events. Consider, for example, the following probabilities derived from Table 3 by the use of Rule 3:

TABLE 8

Demand	Probability
2 or 3	.55
3 or more	.50

If we now want the probability of demand for 2 or more units, it would not do to add .55 and .50 to get 1.05. The difficulty is that the events demand for 2 or 3 units and demand for 3 or more units are not mutually exclusive. They contain the event demand for 3 units in common, and therefore the probability of this event is double-counted. The correct probability of demand for 2 or more units can be obtained either by: (1) applying Rule 3 to the mutually exclusive events listed in Table 3, giving a probability of .80; or (2) subtracting the probability of demand for 3 units from the total of Table 8, thereby eliminating the double-counting and also giving a probability of .80.

Probabilities based on relative frequencies

Perhaps the retailer in the example we are discussing has kept a record of the actual number of units demanded over the past 100 selling days. (Note that to do so he had to keep a record not only of actual sales but also of the number of customers turned away because of insufficient stock on hand.) The results he has obtained are summarized in Table 9.

TABLE 9
Demand History for Past 100 Days

Demand	Number of Occurrences	Relative Frequency
0	5	.05
1	15	.15
2	30	.30
3	25	.25
4	15	.15
5 or more	10	.10
	100	1.00

We observe that the relative frequencies in Table 9, and in fact any relative frequencies, obey the following rules:

Rule 1. Each relative frequency is a number greater than or equal to 0 assigned to an event.

Rule 2. The sum of the relative frequencies assigned to a set of mutually exclusive and collectively exhaustive events is 1.

Rule 3. The relative frequency of an event which is composed of a group of mutually exclusive and collectively exhaustive events is the sum of their relative frequencies.

Except for the substitution of "relative frequency" for "probability," these rules are identical with those given for probability in the last section. As a result, the retailer might be willing to use the historical relative frequencies as his probabilities for deciding on future actions. He could argue, for example, that he considers the event demand for 3 units as being five times as likely as demand for 0 units because he has observed it five times as often in the past.

Another argument which might appeal to the retailer is based on the concept of expected monetary value. Suppose that the retailer had stocked four units on each of the past 100 days. In Table 5 we showed that this stock level led to an expected monetary value of $8.50 when used with probability weights corresponding to the relative frequencies of Table 9, and that this was the highest expected monetary value attainable. In Table 10 we calculate the profitability of such a stock level over the 100-day period.

Thus by stocking four units daily, the retailer would have made a total gross profit of $850 over the past 100 days, or an average of $8.50 per day. Both this total and this average are greater than for any other act. One justification for basing probability weights on historical relative frequencies, therefore, is that this procedure leads to choosing the act which would have been best in the period from which the frequencies were taken. This justification also points out a limitation in the use of observed relative frequencies. They are fundamentally historical data,

TABLE 10
Total Profitability of Stocking Four Units

Demand	Conditional Profit	Absolute Frequency	Profit Times Frequency
0	−$ 4	5	−$ 20
1	+ 1	15	+ 15
2	+ 6	30	+ 180
3	+ 11	25	+ 275
4	+ 16	15	+ 240
5 or more	+ 16	10	+ 160
		100	+$850

and they are only relevant in making future decisions if the user is willing to make the assumption that the future will be like the past. The retailer, for example, may have just recently started to advertise the product he is selling and may believe that future demand will be greater than past demand. Even when relevant historical relative frequencies are available, the decision-maker may be unwilling to base his probability assignments solely on these frequencies. Suppose, for example, that only one selling day has elasped since the retailer began his advertising campaign and on this day five units were demanded. Should the retailer conclude on the basis of this evidence that demand for 5 units has a probability of 1.00, i.e., is certain? Common sense tells us that the answer should be "No"; observing only a single day's demand can hardly enable us to make a perfect prediction of future demands.

Judgmental weights

Even when a decision-maker lacks a firm quantitative base, in the form of relative frequencies, for assigning probabilities to events, he may have a certain amount of qualitative experience which enables him to make such judgments as "event A is more likely than event B." If these judgments could be quantified as numerical probabilities, then these probabilities could be used in calculating expected monetary values. In this section we will demonstrate how this can be done.

Reducing judgments about the likelihood of events to quantitative terms is a problem in measurement. It might help the student in understanding the procedure we are about to outline if we draw an analogy with a more familiar type of measurement. The width of this page, for example, is a quality which is susceptible to quantitative measurement. The way in which we would measure this width is to match the width of the page up against a ruler or yardstick which has previously been calibrated with certain units of measurement—say inches. The numerical width of this page is then taken as the number of units on the standard yardstick which match up against the page.

In order to "measure" probability judgments, we need some sort of "yardstick" against which to match up events. We want our measurements to be such that expected monetary values can be derived from them. For such a yardstick we will use a hypothetical urn into which has been placed 100 hypothetical, serially-numbered balls (i.e., there is a ball 1, ball 2, etc., up to ball 100.) The balls have been thoroughly mixed in the urn. Now suppose that you have been offered the following opportunity: You will call out a number from 1 to 100 inclusive and then will be permitted to draw a ball from the urn without looking. If the number you call matches the number of the ball you draw, you will receive a valuable prize. Otherwise you will get nothing. Suppose that under these circumstances you would feel completely indifferent as to what number to call. For purposes of making this decision, you consider the 100 possible events resulting from drawing from the urn to be equally likely. If this judgment is to be expressed by a numerical probability, then the probabilities assigned to each event must be equal. But since, by Rule 2, the sum of these 100 probabilities must be 1.00, the probability assigned to each event must be .01. This gives us a unit of measurement for our yardstick: the probability that any one ball will be drawn from the urn. Using Rule 3, we can build up our yardstick. For example, the probability of drawing a ball with a number of 1 through 5 inclusive must be .05, since this event is composed of five mutually exclusive events each with probability .01.

Now that we have calibrated our probability yardstick, let us illustrate how it could be used to assign numerical probabilities in a real problem. We will take the event demand for 3 units in the retailer problem as an example. Suppose that the retailer has been offered a valuable prize conditional on one or the other of the two following events:

1. Demand for 3 units.
2. Drawing a ball numbered 1 through 25 inclusive from the urn.

The retailer can choose the event he prefers, but if the event he picks fails to occur he gets nothing.

Now the retailer might be of one of three frames of mind on this choice. First, he might prefer to let his prize depend upon the event demand for 3 units. That is, in his judgment demand for 3 units is more likely than drawing one of the specified balls. Alternatively, he might prefer the drawing, feeling that this event is the more likely. The third possibility is that the retailer may be indifferent between the two events, considering them equally likely. In this case they must each be assigned the same probability. But we have already seen that our probability rules require us to assign a probability of .25 to the event that one of the specified balls is drawn. Thus, in this case the retailer has succeeded in "matching up" the event demand for 3 units with a yardstick event whose probability is given, thereby arriving at a probability assignment to demand for 3 units of .25. If the two events fail to match up, so that one is preferred to the other, the yardstick event can be adjusted by changing the number of balls of which it is composed. In theory, the yardstick could be refined by using more than 100 balls. An urn with 1,000 balls, for example, could be calibrated to probabilities of .001 rather than .01.

We might wonder what is the value of the hypothetical yardstick urn if the probabilities are to be arrived at by judgment anyway. Would it not be possible to write down the probabilities directly without going through the urn procedure? The answer to this question is "Yes," just as it is possible to write down an estimate of the width of this page without using a ruler. Proficiency at estimating widths, however, comes as a result of experience with actual measurements. Similarly, use of the yardstick urn as a hypothetical frame of reference will be helpful in developing one's intuition about probability estimates.

The value of information

One of the merits of the expected monetary value criterion is that it provides a means of determining the value of additional information. This determination is especially useful when additional information can be acquired at a cost: it will pay to acquire the information if an only if its value exceeds its cost. In this part we will show how to calculate the *expected value of perfect information* (EVPI), which is defined by:

> *Expected value of perfect information:* The difference between the expected profit (or cost) of the optimal act under uncertainty and the expected profit (or cost) of the optimal act given complete certainty.

In other words, EVPI is the amount by which expected profit could be increased (or expected cost decreased) by the availability of a perfect forecast.

The benchmark for measuring EVPI is the profit (or cost) which would be experienced if a perfect forecast were available and action were taken accordingly. This benchmark is called *expected profit (or cost) with perfect information.* To illustrate this concept, let us return to the retailer's stock-level problem. Suppose that he were given a perfect forecast of his demand before he placed his order for the day. What would his profit be? A single, definite answer cannot be given to this question since, under uncertainty, we do not know what this forecast will be. We can, however, give conditional answers; that is, we can tell what the profit will be for any given forecast. If, for example, the forecast is that demand will be for 3 units, the retailers' best act will be to stock 3 units and this will give him a profit of $12 (see Table 2). We will call this profit the *conditional profit with perfect information* for a demand of 3 units. (In other problems, of course, we might have conditional cost with perfect information.) In a similar manner, we can get the conditional profit with perfect information for all other possible forecasts of demand. These figures are given in the column of Table 11 headed "Conditional Profit."

Even before looking at Table 11, the reader may have guessed what the next step would be: to get the expected profit with perfect information, we must multiply the conditional profits under certainty by the appropriate probabilities and add. These calculations are also indicated in Table 11. It will be noted that the probabilities assigned to the various possible forecasts are the same as those assigned to the corresponding demands in Table 3. This is because we are assum-

TABLE 11
Calculation of Expected Profit Under Certainty

Forecast	Conditional Profit	Probability	Product
0	$ 0	.05	$ .00
1	4	.15	.60
2	8	.30	2.40
3	12	.25	3.00
4	16	.15	2.40
5 or more	20	.10	2.00
		1.00	$10.40

ing perfect forecasts. A forecasted demand of three units, for example, would be perfect if and only if actual demand is for three units, and this demand has a probability of .25.

According to Table 11, the expected profit with perfect information in the retailer problem is $10.40. The optimal act under uncertainty was shown to be stock 4 units in Table 5, and it has an expected profit of $8.50. Hence we can calculate EVPI as follows:

Expected profit with perfect information	$10.40
Expected profit under uncertainty	8.50
Expected value of perfect information	$ 1.90

An alternative method of calculating EVPI is illustrated in Table 12.

TABLE 12
Alternative Calculation of EVPI

Demand	Probability	Conditional Profit Stock 4	Perf. Inf.	CVPI	EVPI
0	.05	−$ 4	$ 0	$4	$ .20
1	.15	+ 1	4	3	.45
2	.30	6	8	2	.60
3	.25	11	12	1	.25
4	.15	16	16	0	.00
5 or more	.10	16	20	4	.40
	1.00				$1.90

In Table 12, the column headed "CVPI" gives the Conditional Value of Perfect Information, which is simply the difference between conditional profit with perfect information and conditional profit with a stock level of 4, the optimal act under uncertainty. If, for example, the retailer gets a perfect forecast that demand will be for 3 units, he will be able to improve his profit by $1 (by stocking one less units than the 4 required under uncertainty).

In order to calculate the EVPI in a given decision problem by either of the methods just presented, it is first necessary to determine the optimal act under uncertainty. We will now illustrate another method of analysis in which the optimal act and the EVPI are determined simultaneously. To calculate EVPI, we use the profit (or cost) of action with perfect information as a reference point against which we compare the profit (or cost) of the optimal act under uncertainty. The EVPI of $1.90 calculated above, for example, indicates that the retailer's best stock level under uncertainty is expected to be $1.90 less profitable than action with perfect information. Now we could also compare other acts with action with perfect information. In Table 5, for example, it is indicated that a stock level of 3 units has an expected profit of $8.25. This is $2.15 less than the expected profit with perfect information of $10.40 given in Table 11. We can calculate this $2.15 figure in a second way, analogous to the second way of calculating EVPI given above. But first we must define the

Conditional opportunity loss of an act given an event: The difference between the conditional profit (or cost) of the act and the conditional profit (or cost) with perfect information.

Conditional opportunity loss is always taken as being positive or zero. That is, it is calculated as either: (1) conditional profit with perfect information minus conditional profit of the act; or (2) conditional cost of the act minus conditional cost with perfect information. To illustrate the idea of conditional opportunity loss, the losses for the retailer example are given in Table 13; a table of this sort is called a *loss table*. The student should check a few of the values in the table by applying the above definition to the conditional profits given in Tables 2 and 11.

TABLE 13
Loss Table for Retailer Example

Demand	Act (Stock Level)					
	0	1	2	3	4	5
0	$ 0	$ 1	$ 2	$3	$4	$5
1	4	0	1	2	3	4
2	8	4	0	1	2	3
3	12	8	4	0	1	2
4	16	12	8	4	0	1
5 or more	20	16	12	8	4	0

To give some meaning to the term conditional opportunity loss, let us look at the losses corresponding to a stock level of three units. If this act is chosen and actual demand turns out to be three units, the retailer will make the maximum profit he could have made given that demand; he has lost nothing. If, on the other hand, actual demand is for four units, then by stocking only three units the retailer has lost the opportunity of selling a fourth unit on which his gross profit would have been $5 − $1 = $4; this $4 is then his conditional opportunity loss given a demand for four units. Finally, if demand is for only two units, the retailer will be overstocked by one unit. He has therefore lost the opportunity of saving the $1 cost of this unit by failing to order it.

The loss table gives conditional values only. In order to determine the optimal act, we must find the *expected opportunity loss* (EOL) of each act, i.e., the weighted average of the conditional opportunity losses using the probabilities as weights. The best act will then be the one with the lowest expected opportunity loss. The expected opportunity losses corresponding to Table 13 are given in Table 14. The student should check the calculation of at least one of these figures.

TABLE 14
Expected Opportunity Losses of All Possible Stock Levels

Stock Level	Expected Opportunity Loss
0	$10.40
1	6.65
2	3.65
3	2.15
4	1.90
5	2.40

Several facts may now be observed about *expected opportunity loss* (EOL):

1. The EOL of action with perfect information is, of course, always $0, since it is the reference point from which opportunity loss is measured.
2. The EOL of the optimal act under uncertainty is also the expected value of perfect information—in this case $1.90. The reason for this can be seen by comparing the definitions of conditional value of perfect information and conditional opportunity loss; for the optimal act these definitions are identical. The optimal act is the closest we can get to 0 opportunity loss on the basis of available information.
3. The EOL of all acts other than the optimal one are greater than the EOL of the optimal act. By comparing Tables 5 and 14, we can see that the differences in EOL between any two acts is equal to the difference between their expected profits. A corresponding statement would be true if we were working with cost rather than profits.

Thus, the use of expected opportunity loss as a criterion enables us simultaneously to: (1) find the optimal act; and (2) find the EVPI, which is the EOL of the optimal act.

Expected utility and attitude toward risk

On page 121, we pointed out that circumstances exist in which a decision-maker could feel that use of the expected monetary value criterion was inappropriate. These circumstances occur when one act has a higher expected monetary value than a second but also runs a substantially greater risk. In such a case, the decision-maker might well decide that the additional expected value of the first act is not sufficient to compensate him for its additional risk.

Previously we illustrated this point with a problem in which you were asked to choose between two acts: one leading to a "profit" of $0 with absolute certainty, and the other leading to a 50-50 chance at a profit of +$55,000 or −$45,000. While the second act has an expected monetary value of $5,000, you might still prefer the first act because you wish to avoid the risk of substantial loss in the second. The amount of risk you are willing to assume is a matter of personal preference.

But while attitude toward risk is a matter of personal preference, this does not imply that it cannot be dealt with quantitatively. We have already shown how one subjective factor—relative likelihood of events—could be expressed numerically through the use of a "yardstick" urn. Somewhat more surprisingly, a very similar device can be used to reduce attitude toward risk to a quantitative basis!

Essentially, the device is to use a "yardstick" urn to assign conditional values to acts given events in such a way that the weighted averages of these conditional values are valid guides to action, even if the weighted average of conditional monetary values is not. The conditional values we will obtain are called *utilities,* and the criterion which results is called the *expected utility criterion.* Those who are familiar with the concept of "utility" as economists often use it, should be warned that it is not possible to take the weighted average of any arbitrary set of utility numbers as a guide to action. The utility numbers must be derived by a procedure which explicitly recognizes differences in riskiness.

A 50-50 chance at either $55,000 or −$45,000 might not be worth $5,000 to you, or even $0. Still, there is probably *some* amount which this chance *is* worth. Suppose, for example, that the alternative to the 50-50 chance was a certain loss of $3,750. Under these circumstances, you might very well be indifferent between the 50-50 chance and the certain loss of $3,750. We could then say that these two acts are of equivalent value and that, therefore, the 50-50 chance is "worth" −$3,750.

Notice that this assignment of value to the 50-50 chance is entirely subjective, which is as it should be if it is to reflect personal attitude toward risk. There is no logical principle to say that the 50-50 chance is "worth" −$3,750, but only personal preference. Because of this, the assignment of a value to the 50-50 chance does not, in and of itself, help in making a better decision; it only reflects how the decision-maker would decide based on personal preferences.

Not all decisions are so easy to resolve judgmentally. It would not be so easy, for example, to assign a subjective value to the act described by Table 15. While this act has a lower expected monetary value than the 50-50 chance, intuitively it seems less risky. Our objective will be to show how choices in simple situations can be combined logically to aid the decision-maker in evaluating complex acts as this.

Instead of saying that a 50-50 chance at $55,000 or −$45,000 is "worth" −3,750, we could just as well turn this statement around and say that a certain loss

TABLE 15

Description of a More Complex Act

Consequence	Probability
$55,000	.10
25,000	.25
0	.40
−25,000	.15
−45,000	.10
EMV = $ 3,500	1.00

of $3,750 is "worth" a 50-50 chance at $55,000 or −$45,000. While this approach may seem less natural, since we are accustomed to thinking of value in monetary terms, it will turn out to be more useful in accomplishing our purpose.

Having shown how judgment can be used to determine that two acts have the same "value," let us now use this process to set up a "yardstick" for assigning utility values to consequences. As our "yardstick," we will use a hypothetical urn containing a number of balls, some of them marked "$55,000" and the rest marked "$45,000." You will be permitted to draw a ball unseen from the urn, and you will then receive or be required to pay the amount indicated on the ball. Before we can use this "yardstick" urn to measure utility, we will have to calibrate it, i.e., agree on a unit of measurement. Clearly, the higher the proportion of balls in the urn marked "$55,000," the more valuable the option of drawing from it. This suggests that we use the proportion of balls so marked as an index of utility value. Thus, if the urn contains only balls marked "$55,000," so that the probability of winning $55,000 is 1.00, we would say that it represented a utility value of 1.00; while if it contains no such balls, we would say that its utility value was 0.00. This assignment of values is admittedly somewhat arbitrary but will not detract from the validity of the yardstick. The situation is analogous to the measurement of temperature. In calibrating a Fahrenheit thermometer, the freezing point of water is arbitrarily assigned a value of 32° and the boiling point a value of 212°, while on a Celsius (centigrade) thermometer these points are arbitrarily assigned values of 0° and 100° respectively. Despite the arbitrariness of these assignments, both temperature scales are perfectly valid and convey the same information.

To illustrate the use of the "yardstick" urn, let us now measure the utility value of a loss of $3,750. Assuming, as we did earlier, that you would as soon incur this loss as take a 50-50 chance at $55,000 or −$45,000, these two alternatives must be assigned the same value. But, according to the way we have calibrated our "yardstick," the 50-50 chance is assigned a utility value of 0.50. Since a loss of $3,750 must be assigned the same value, its value must also be 0.50.

As a further illustration, let us assign a utility value to $0. Evidently, given that you prefer $0 for certain to the 50-50 chance, $0 is "worth" more than the 50-50 chance and so its utility value must be more than 0.50. But how much more? If we represent the 50-50 chance by an urn containing 50 balls marked "$55,000" and 50 marked "−$45,000," we can increase the value of the urn by substituting "$55,000" balls for "−$45,000" balls. Eventually we may expect to be able to construct a "yardstick" urn which "matches up" against $0 for certain in the sense that you would find them equally attractive. (Here, as in the case of probability measurements, it may theoretically be necessary to increase the refinement of the measuring process by using an urn containing more than 100 balls, but this is not of great practical concern.) If this point is reached when the "yardstick" urn has, say, a 54-46 split, then we must assign a utility value of 0.54 to $0.

Proceeding in the same way, we could find the utility of any other monetary value between −$45,000 and $55,000. Of course, use of the "yardstick" urn to find the utility of every possible monetary value in every decision problem in which risk is an important factor would be so time-consuming as to be impractical. Fortunately this difficulty can be side-stepped in a fairly simple way. The decision-maker can determine the utility of a few selected monetary values, plot these values on a sheet of graph paper, and then fit a smooth curve to the plotted points. The curve can then be used to find by interpolation utilities for additional monetary values. Such a curve is illustrated in Figure 13.1.

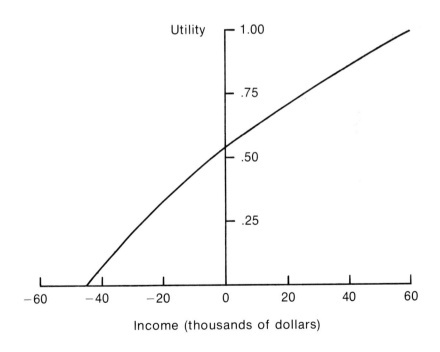

Figure 13.1 Utility curve

So far we have shown how to use a "yardstick" urn to measure the utility of any given monetary payment. This procedure does not in and of itself, however, help the decision-maker to make better decisions, since the numerical measurements do no more than reflect how he would act in simple choice situations on the basis of his personal preferences. The numbers do not tell him how he *should* behave in these situations but only how he *would* behave. Once again it might help to draw an analogy with a more familiar type of measurement, as we did previously in this chapter. In measuring the width of this page with an ordinary yardstick, we could obtain a value of approximately 6 inches meaning that the width of this page matches up against 6 inch-units on the yard-stick. Similarly, the value obtained by measuring the length of this page would be approximately 9 inches. In and of themselves, these values convey only the information that the page is longer than it is wide; the number of length-units exceeds the number of width-units. By applying the rules of arithmetic to these measurements, however, we can derive new conclusions without having to use a yardstick directly. For example, we can find the width of a double-page spread: it is 12 inches obtained by addition of 6 inches and 6 inches. Thus a double-page spread is wider than it is long. We could, of course, verify this conclusion empirically by using a yardstick, but we do not need to rely on the yardstick. In short, *direct* measurements of length (by means of a yardstick) plus the rules of arithmetic enable us to arrive at *indirect* measurements of length.

In a similar way, *direct* measurements of utility (by means of a yardstick) plus the rules of arithmetic enable us to arrive at *indirect* measurements of utility. In particular, we can find the utility value of any act by:

1. Determining the utility value of each consequence of the act by direct measurement.
2. Taking the weighted average of these utility values, using the appropriate probabilities as weights.

The resulting number is called the *expected utility* of the act. The expected utility of the act described in Table 15, for example, is calculated to be 0.545 in Table 16. The utility values used in this table have been read from Figure 13.1. That figure in turn, you are reminded, has (by assumption) been constructed from direct measurements using a "yardstick" urn.

TABLE 16
Calculation of Expected Utility

Consequence	Utility Value	Probability	Weighted Utility
$55,000	1.00	.10	0.100
25,000	.76	.25	.190
0	.54	.40	.216
−25,000	.26	.15	.039
−45,000	.00	.10	.000
		1.00	0.545

Since the expected utility of this act is 0.545, it is supposedly worth more than a 50-50 chance at $55,000 or −$45,000, which has a utility value of only 0.50. The reader may well be skeptical of this conclusion, though, and so let us see whether we can verify it in another way, using direct measurement (just as we could verify indirect length measurements with a yardstick).

We can represent the act described by Table 15 by an urn containing 1,000 balls labeled as follows:

Label	No. of Balls
$55,000	100
25,000	250
0	400
−25,000	150
−45,000	100
	1,000

The balls in this urn will be thoroughly mixed and you will be allowed to draw one ball. You will then be given the amount of money corresponding to the label on the ball you have chosen. If you are willing to assume that each ball in this urn is equally likely to be drawn, then it is clear that the urn is equivalent to the act described in Table 15. Now let us bring our "yardstick" urn into the problem. To start with, we will use the "yardstick" to measure the utility of $0. In our earlier illustrations, we have assumed that you would be indifferent between $0 for certain and a 54-46 chance at $55,000 or −$45,000. This being the case, we can put 400 balls into the "yardstick" urn of which 54%, or 216, are labeled "$55,000" and 46%, or 184, are labeled "−$45,000." Since the chance of drawing from this "yardstick" urn is worth exactly the same to you as $0 for certain, let us next *replace* the 400 balls labeled "$0" in the original urn with the 400 balls in the "yardstick" urn. We will assume that this replacement will not affect the value of drawing from the original urn, since it involves only the exchange of two things of equal value. The composition of the original urn will then be:

Label	No. of Balls
$55,000	316
25,000	250
−25,000	150
−45,000	284
	1,000

We can next perform the same operation on the balls labeled "$25,000." In Table 16, we assumed that the utility to you of $25,000, as obtained by direct measurement, was 0.76. Therefore, $25,000 is "worth" as much as the opportunity of drawing from a "yardstick" urn containing 250 balls, of which 76%, or 190, are labeled "$55,000" and 24%, or 60, are labeled "−$45,000." Substituting those

250 balls for the 250 balls labeled "$25,000" in the original urn, the composition of that urn becomes:

Label	No. of Balls
$55,000	506
−25,000	150
−45,000	344
	1,000

Finally, repeating the process with −$25,000 we can substitute for the 150 balls labeled "−$25,000" in the original urn the contents of a "yardstick" urn containing 39 (26%) balls labeled "55,000" and 111 (74%) balls labeled "−45,000." As a result, the original urn is now composed as follows:

Label	No. of Balls
$55,000	545
−45,000	455
	1,000

Since the original urn is now composed of "$55,000" and "−$45,000" balls only, it is itself a "yardstick" and we can use it to make a direct measurement of the utility of the act described in Table 15. This direct measurement, 0.545, agrees with the indirect result obtained in Table 16, thus verifying the validity of the indirect procedure.

The expected monetary value criterion discussed in this chapter is an appropriate criterion only when risk is not an important consideration in comparing acts. When risk is important, the concept of expected value can still be used, except that now it is necessary to measure the value of consequences on a "utility" scale instead of in ordinary monetary terms. The resulting criterion is called "expected utility." We have shown how a utility scale can be constructed and demonstrated its validity.

CHAPTER 14

BACKGROUND IN LANGUAGE

The language of Wall Street

Often you hear that the mechanics of the stock market are simple once you learn the language. It is said that many people do not buy stocks since they think investing is a complicated business, mainly because a lot of unfamiliar words are used. As a result many books attempt to explain the meaning of the technical words by describing the things that they stand for. Words such as investment, common stocks, preferred stocks, bonds, new issues, stock exchanges, over-the-counter, and so on are explained in detail. However, the important words, that is, the only words that can actually describe the intrinsic behavior of the stock market, are never described and in fact are rarely used on Wall Street.

What are these important words? They are the words that make up the language of the science of probability. Although Wall Street makes a concerted effort to exclude these words from its language, it is not entirely possible to do so. The reason is that the concept of probability is very deeply embedded in human experience, and cannot be obliterated by the deliberate efforts of any special-interest group.

For example, let us look at the manifestation of probability concepts in the English language.

The English word *hap* originates in the Old Norse word *happ*. The noun *hap* means "luck, chance," whereas the verb *to hap* means "to befall." From this basic word the following English words are derived.

(1)	*to happen* (v.i.):	to occur by chance
(2)	*happening* (n.):	occurrence, that which happens
(3)	*haply* (adv.):	by chance
(4)	*haphazard* (adj.):	random, determined by chance
(5)	*perhaps* (adv.):	by some chance, maybe
(6)	*happy* (adj.):	favored by luck or fortune
(7)	*happily* (adv.):	by good fortune
(8)	*happiness* (n.):	good luck, a state of well-being
(9)	*happy-go-lucky* (adj.):	trusting to luck, easygoing
(10)	*hapless* (adj.):	unlucky
(11)	*mishap* (n.):	ill luck, an injurious accident
(12)	*happenstance* (n.):	a circumstance due to chance
(13)	*mayhap* (adv.):	by some chance, maybe

A typical report on a stock in the language of Wall Street would read as follows:

> "Company XYZ has established an excellent record of earnings growth for the past three years. However this year's results were very disappointing but prospects for the coming year are encouraging. The company has greatly expanded its position in computer peripheral equipment which together with pending research efforts in related areas augur well for the continued growth of the company's operations. In our opinion this is a reasonably valued situation which can be accumulated by accounts desiring long term appreciation potential."

The above report is almost completely lacking in probability concepts. Let us now rewrite the same report by going to the opposite extreme:

> "Although Company XYZ happened to have an excellent record of earnings growth three years in a row, this year's results were a mishap, but perhaps next year things will be better. The company has greatly expanded its position in computer peripheral equipment which together with its haphazard research efforts might produce a happy result. In our opinion this is a situation which can be accumulated by accounts willing to bet their money and happiness on the happenstance that this presently hapless company happily will again prosper."

The nouns

hap	*luck*	*chance*
fortune	*lot*	*hazard*

cannot be properly defined. They are more or less synonymous. They all agree in designating "that which *happens,* that which befalls either partially or entirely as the result of unknown, unconsidered, or unpredictable forces." Synonymous words such as these continually grope their way through language by connotation, that is by their suggestive significance. There is nothing less precise than connotation, which changes from person to person and from time to time. It is based on associations, many of which are below the conscious level. Nevertheless, without connotation we could not have language as we know it.

The usage of such words is illustrated in the following passage, an episode about a Wall Street specialist named Smith.

> "After a cold morning shower, Smith took the busy subway to the Stock Exchange. Since the day before he had lost the mood of the market. He needed to reestablish that focus which is half intuitive and which, with a slow pulse and a cool temperament, Smith knew to be the essential equipment of any gambler set on winning. Smith had always been a gambler. He loved the buy and sell figures contained in his specialist book and the feel of the drama of the people around his trading post. He was amused by the impartiality of the market and its eternal bias. He liked being a speculator and from his post

to take part in other investors' dramas and decisions, until it came to his turn to make that vital "buy" or "sell" decision generally on a better than fifty-fifty chance. Above all, Smith knew that everything was one's own fault. On Wall Street there is only oneself to praise or blame, for luck is a servant and not a master. Luck has to be accepted with a shrug or taken advantage of all the way. But luck has to be understood and recognized for what it is and not confused with a faulty appreciation of the odds, for, on Wall Street, the worst mistake is to confuse bad timing for bad luck. And luck in all its moods has to be loved and not feared. One day, Smith knew, he could be wiped out by a string of bad luck. When that happened he realized that he would be branded as a loser,—the result he saw so often in others,—the loss of all your money, the acceptance of fallibility. At that point he would have to turn to his colleagues—the other exchange members—to bail him out."

Probability concepts in everyday language

The noun *luck* usually refers to that which happens to one personally or individually. The word *luck* is associated etymologically and by continual use with gambling, as

The computer stocks have had a wonderful run of luck.
Don't abuse your luck by investing in too many different issues.
Playing the stock market is strictly luck and nothing else.
If he couldn't cover his short position, it would be a kick in the teeth to the luck which had been given him.
Luck only enters the bond market at sporadic intervals.
Luck has no influence in the long run over the inflationary policy of the government.

The word *luck* is used constantly in colloquial speech, as

His profit on that investment was just dumb luck.
It was just his luck to buy that stock.
He had had luck in the market today.
The newcomer had good luck.
The man tried his luck on convertible securities.

The word *luck* unqualified can imply a happy outcome, as

Wish me luck on these hundred shares.
He had luck in airline stocks.

The derived word *lucky* (adj.) means favored by luck, fortunate, as

My lucky stock is Natomas.
I have no lucky stocks.
He was lucky to take profit again.
You were lucky to find that new issue.
Lucky in cards, unlucky in love.

The English word *chance* (n.) comes from the Latin *cadentia* which means "a falling, especially of dice or of fortune." The Latin word *cadentia* itself is derived from the Latin word *cadere* meaning "to fall, to happen." Thus we have found an important link: the word *chance* has its roots with the falling of dice.

Consequently, the noun *chance* serves often as the general name for the incalculable and fortuitous element in human existence and in nature as

> Most share prices are influenced by chance.

In common usage, *chance* seldom loses implications derived from its original association with the throwing of dice and the selection of one outcome out of many possible outcomes by this means. Consequently: It may mean "determination by irrational, uncontrollable forces," as

> He left things to chance when it came to investing his life savings.
> I am not a gambler; I like to play safe, to make certain, to leave as little as possible to chance in my investments.

It may mean any one of the contingencies on which a player takes a risk in a game of chance, or more generally any risk or gamble, as

> Computer stocks are my chance.
> Take a chance and accept a long position in oils.
> It is my only chance, so sell.
> Sell because it's the last day before a long weekend and I don't want to take any chances on news that might happen over the weekend.

It may mean an opportunity that comes seemingly by luck or accident, as

> When I get a chance, I'll buy Standard at 70.
> It was my last chance to sell before it peaked.
> He has long hoped for a chance of dumping that issue at a profit.

It may mean a possibility of something happening, as

> I'll have a chance to buy some GM later.
> There is little chance of the stock breaking out of that pattern.
> The chances are even for a bear market in view of a tighter monetary policy.
> Do you think you have a good chance to sell before the bad news comes out?

It may mean a possibility of success among many possibilities of failure, as

> He is always willing to take a chance on the market.
> What are his chances in gold mining stocks?
> The chances are against him.
> He has one chance out of a million.
> The crash happened so fast that he didn't have a chance.

It may mean a measure that an event happens, as

> What is the chance of the Dow Jones Industrial Average hitting the previous high?

Its chance of happening is 2 out of 10, or 0.2.
What is the chance of drawing two kings in a poker hand?
What chance of a rally this week?
There is little chance.

The English word *chance* (v.i.) means to happen, come, or arrive without design or expectation, as

He chanced upon this new concept stock.

Chance (v.t.) means to take the chances of, to risk, as

To chance buying in a bear market is foolish.

The noun *case* is etymologically related to *chance*. *Case* comes from the Old French *cas*, which is from the Latin *casus*, which in turn is from the Latin *cadere* meaning to fall, to happen.

The adjective *chance* means happening by chance. Synonymous words are *haply, accidental, fortuitous,* and *aleatory*.

A word related to *chance* is *accident*. The word *accident* came from the Latin word *accidere* which means "to happen" and which is a compound of *ad* plus *cadere* (to fall). Thus an accident is a happening or an event which takes place without one's foresight or expectation, and now often means one of an afflictive, injurious, or unfavorable character, as

One never speaks of "accidents" in the stock market.

The derived word *accidental* (adj.) means "happening by chance or unexpectedly." Although the usage of the word accidental usually stresses chance, it now sometimes may stress nonessentiality.

The word *fortune* (n.) came from Old French *fortune* which in turn came from Latin *fortuna*. In addition to its meaning of "luck, chance," it also means (sometimes capitalized) the personified power of chance, the personification of the cause of that which befalls in a sudden or unexpected manner. Thus we hear about *Dame Fortune* who is the same mysterious woman as *Lady Luck*. Sometimes she is pictured as having a large wheel which she spins to determine the various happenings of life.

The derived word *fortunate* (adj.) means "coming by good luck, auspicious, lucky." The synonyms *lucky, fortunate,* and *happy* all connote "having a favorable issue." *Lucky* implies success by chance rather than as the result of merit. *Fortunate* is less suggestive of a favorable accident and may carry the connotation of being watched over by the higher power or of being favored beyond one's deserts. *Happy* combines the implications of *lucky* and *fortunate* to express gratification in the sense of well-being and of complete satisfaction. Thus *happy* becomes synonymous with *glad* (from Anglo-Saxon *glaed* meaning "bright, glad") to mean "characterized by joy or pleasure, pleased." The English word *glad* is the same as the Swedish word *glad*, both originating from the same Teutonic word. Whereas in English we have seen that *happy* has become synonymous with *glad,* in Swedish it has turned out that *lycklig* has become synonymous with *glad.*

The English word *fortuitous* (adj.) comes from the Latin word *fortuitus,* which in turn is from the Latin *fors,* meaning "chance." *Fortuitous* means "happening by chance or accident" and is synonymous with *accidental.* Whereas both *accidental* and *fortuitous* mean "not expected, outside of the regular course of things," *fortuitous* more strongly suggests chance than accidental and often connotes absence of a cause.

The word *lot* comes from the Anglo-Saxon word *hlot,* which was an object used as a counter or check in determining a question by chance. *To choose by lot* is the use of such a counter as a means of deciding something. *Lot* has come to mean "that which befalls one from a choice by lot, and hence a share of allotment." *Lot* has also come to mean "hazard, fortune, especially the fate which falls to one by the will of an overruling power," as

The small investor is a man content with his lot.

In the usage of *lot* to mean "the state or end predetermined for one," there is always the suggestion of the operation of blind chance. Like the work *luck,* the word *lot* usually refers to that which happens to a person as an individual.

The derived word *lottery* (n.) means a scheme for the distribution of prizes by lot, especially such a scheme in which lots, or chances, are sold, and has come to mean figuratively an affair of chance.

The word *hazard* (n.) comes from Arabic *al-zahr* which means "the die." Originally *hazard* was a dice game. Thus hazard came to mean "chance." In addition hazard also has come to mean "risk, danger, peril."

The derived word *hazardous* (adj.) thus means both "depending upon chance or luck" and "dangerous, risky." In fact *hazardous* implies so many chances of evil and harm that the thing so described is exceedingly dangerous.

The Latin word *alea* meaning "die, chance" did not come directly into the English language. Nevertheless, this word is known from the famous expression of Caesar when he crossed the Rubicon River:

Alea jacta est (which means *the die is cast*).

The derived latin word *aleatorius* has come into English as *aleatory* (adj.). It means "pertaining to or resulting from luck." In legal terminology, it means "depending upon an uncertain event or contingency as to both profit and loss." Thus *aleatory contracts* include lottery agreements, wagering contracts, and insurance contracts.

Let us now look at the word *random.* Its earliest known meaning is that of furious action, such as a charge of cavalry. It seems to be connected with the Teutonic word *rand,* which means *brim,* and implies the furious and irregular action of a river full to the brim. The English word random is related to the Old French *randon,* which means "violence, rapidity." *Random* (n.) means an aimless course or progress. It is used chiefly in the expression *at random,* which means without definite aim, direction, rule, or method.

Random (adj.) and *haphazard* (adj.) are synonyms. Both mean "having a cause or a character that is determined by accident rather than by design."

Random signifies that there is no fixed or clearly defined aim, purpose, or evidence of method or direction, and hence implies little or no guidance by a governing mind or objective, as

> Stock prices drifted at random for the second day with low volume trading.
> Stock prices wandered at random, after a rally effort early in the session, led by glamor-sector favorites, lifted most issues from their lows of the day.

Here random indicates the aimless character of the performance, as contrasted to the definite intention to hit a certain mark.

Haphazard, on the other hand, signifies "an aimlessness or randomness that is more or less at the mercy of chance," as

> His portfolio was a haphazard arrangement of bonds and preferred stocks.
> Weakness in blue chip issues resulted in a haphazard performance for the general market.
> Mutual fund performance is often haphazard in a bear market.
> He didn't like haphazard talk; he wanted a detailed breakdown of what to buy and sell.

In probability theory the word *random* is used as the word *haphazard* is used in ordinary language, that is, with a definite connotation of chance.

Next let us look at the English adjectives *probable* and *likely,* and the corresponding nouns *probability* and *likelihood.*

The Latin adjective *probabilis* comes from the Latin verb *probare* meaning "to try, to test, to prove." The English word *probable* comes through French from the Latin *probabilis*. Thus etymologically *probable* is related to *provable*.

The Latin word *verisimilis* (adj.) comes from the Latin *verus* true (genitive: *veri*) + *similis* like. The following correspondence exists among different languages.

Latin:	verisimulus	(*veri* true + *similis* like)
English:	verisimilar	(*veri* true + *similar* like)
German:	wahrscheinlich	(*wahr* true + *scheinlich* appearing)
Swedish:	sannolik	(*sanno* true + *lik* like)
	trolig	(*tro* true + *lig* like)
Russian:	veroyatnii	(*vernii* true)
French:	vraisemblable	(*vrai* true + *semblable* like)
Italian:	verisimile	(*veri* true + *simile* like)

All the above words were built up from the basic stems shown in the parentheses to express the meaning: true-like, likely, true-seeming, true-appearing, true-resembling, true-similar.

The English words *probable* and *likely* are synonyms. They signify "uncertainty that may be, or become, true, real or actual." Something is *probable* if it has so much evidence in its support or seems so reasonable that it commends itself to the mind as worthy of belief, although not to be accepted as a certainty. Thus the *probable* conclusion from evidence at hand is the one which the weight of evidence supports even though it does not provide proof, as

> That fund was the probable buyer of the 10,000 share block.
> The probable cause of the drop was decreased earnings.
> What are the probable expenses for the investment service?
> That is the probable result of such an investment.

The word *likely* is very close to the word probable, and so they can often be used interchangeably. In contrast to *probable, likely* does not as invariably suggest grounds sufficient to warrant a presumption of truth, as

> That is not the probable end of the bull market, but it is still likely.
> It is a time when something really may be happening, or is at least likely to happen.

Also likely is sometimes used in the sense of *promising* because of appearances, ability to win favor, etc., as

> This stock is a likely candidate.

Using these words we may establish a scale of expressions, as

> It is *certain* to happen. (It is *sure* to happen.)
> It is *very likely* to happen.
> It is *likely* to happen.
> It is *as likely* to happen *as not.*
> It is *unlikely* to happen.
> It is *very unlikely* to happen.
> It is *certain not* to happen. (It is *sure not* to happen.)

or as

> This issue is the *certain* winner.
> This issue is the *probable* winner.
> This issue is a *likely* winner.
> This issue is an *unlikely* winner.

The noun *likelihood* is derived from *likely;* the noun *probability* is derived from *probable. Likelihood* and *probability* are synonyms. They mean "quality or state of being likely or probable," as

> The likelihood of winning in the stock market is small.
> The probability of the stock falling is great.
> The probability of making a lot of money is small.

As we have seen, the word *chance* is related etymologically to the falling of a die. On the other hand, the word *probability* is related etymologically to trying or testing a thing against some truth. In mathematical usage, the words *probability* and *chance* overlap completely and signify a "measure that an event happens."

Use of language

The many-fold connotations of words used in every day life as well as in probability theory can cause trouble when these same words are for the most part avoided when it comes to investing in the stock market. George Christoph Lichtenberg (*Vermischte Schriften,* Erster Teil, II, 1, New Edition, Gottingen 1853, Vol. 1, p. 79) has said

"All our philosophy is a correction of the common usage of words."

In this book we do not wish to undertake a correction of common word usage of Wall Street investment houses. Instead we wish to develop certain concepts on a mathematical basis with a view toward their intuitive background and their applications to give insight to the operation of the stock market. The value of these concepts can only be judged from their fruitfulness, both as creative ideas and as useful instruments for providing a more rational and operational basis for realizing profits in the stock market, and for cutting losses. In conclusion the operation of the stock market can be well summed up by the following statement (which ends with the word "chance") by the famous financier Bernard M. Baruch:

"If you are ready and able to give up everything else, and will study the market and every stock listed there as carefully as a medical student studies anatomy, and will glue your nose to the ticker tape at the opening of every day of the year and never take it off till night; if you can do all that, and in addition have the cool nerves of a gambler, the sixth sense of a clairvoyant and the courage of a lion—you have a small chance."

In the next chapter we will see why the insiders constantly feel it necessary to give such statements on why the small investor almost always loses his money. These statements never give the true reasons, either by accident or design.

CHAPTER 15

STOCK MARKET SPECULATION AS A GAME

Model building

Probability theory has a dual scientific role in the analysis of the stock market: one is the search for ways to make or protect money and the other is the research required to develop new methods and better scientific understanding of market action. Progress in either one of these roles is dependent upon the other. The exchange of ideas and the communication needed to attain these goals depends upon the use and development of general scientific models that are broad enough to cover and bridge the gap between these two roles. Perhaps the most important thing that must be recognized and honored is that it is essential to have a clear understanding of the psychology of the people involved with the stock market, whether they be active traders or dormant investors. In this respect it is necessary that the computer scientist has a broad objective under which he operates, but at the same time he must be given full freedom and opportunity to develop his own creativity and imagination. With these incentives the science of probability can help those who make full use of it.

Let us now look at some of the fundamental notions of model building. First, what are models? A model is simply a representation of something else. Typically, it is a representation in which details that appear unessential for the intended use of the model are omitted. A stock market model is supposed to represent the stock market in certain significant respects.

However, the advantage of a model over the actual stock market is the understanding of the market action made possible by the model. With these purposes in view, it is clear that the objective in model building should not be to create an exact duplicate of the real market. In summary, models are, or should be, useful substitutes for what is modelled.

If, in fact, there exists a useful connection between the behavior of the market and the corresponding behavior of the model, then the model can be useful in analyzing data from the market and making buy, sell, and hold decisions.

Let us now say a few words about how models may be described. Models may be described verbally, graphically, physically, or by mathematical functions and equations. Models may also take the form of computer programs. Some modes of expression might well be more convenient for describing and using some models than would using other modes of expression. However, most models can be expressed in several different modes.

In general it is good to have several expressions of the same model but at different levels of sophistication. The approach of model building is broad and flexible, and a host of types of models can be constructed for a variety of purposes. At the highest level any given model should be connected with other models in order to bring out the interdependences and interrelations. At the intermediate level the model should take into account all the important market parameters, so the user has the flexibility and scope to experiment with and adjust the model to the actual situation. At the deepest level the model should be built on a firm mathematical and physical foundation in order to support the ultimate consequences resulting from the model.

In essence model building is a systematic coordination of theoretical and empirical elements of knowledge into a joint construction. It is important to discern between, on the one hand, the hypothetical assumptions that constitute the theoretical part of the model and, on the other hand, the empirical observations that the model serves to interpret. Empirical observations enter in different ways in the model construction: at an early stage when observations and experience are accumulated, whether a tentative model has or has not been formed; at more advanced stages, in assessing the parameters of the model on the basis of observations and in testing the theoretical model against the empirical evidence.

Because of the large mass of data that must be processed in the analysis of the stock prices and volumes, most market models must have a statistical basis. In the actual operational use of the model, data must be sifted and sorted out. In order to make the correspondence with the actual situation, the research models must also allow for noisy and uncertain situations which means that the models must incorporate statistical parameters. Statistical data and statistical methods are in everyday use in most or all sciences that unravel the world around us, and statistical techniques have proven indispensable in a great many areas of human activity. In this respect the stock trader historically has been ahead of the research analyst, for the trader must every day come to grasps with the market and make decisions about uncertain events based on his evaluation and assessment of statistical data, whereas all too often the research analyst has confined himself to the narrow bounds of strictly deterministic mathematical models which can never adequately describe the complex factors present in the market. However, in trading as in any other branch of human activity scientific method is synonymous with the methods of model building, or to put it otherwise accumulated knowledge takes the form of scientific models. With the introduction of large scale digital processing of market data the operative uses of the scientific models have become essential, and as a result the research analysts stand to make significant advances in model building. These advances make current market models useful both in the daily trading and in research for better methods. In this regard market analysts are at the forefront of scientific knowledge today. They strive to coordinate the practical and the theoretical aspects of model building with specific regard to their similarities and differences. A first incentive for an integrated treatment is, of course, the predictive aspects of models which make use of both theoretical and empirical results. The trader when using market data would then try to make use of the predictive aspects as much as possible. There is further a direct need for an integrated treatment because the available market data is usually of a composite nature, partially psychological and partially observational. The great rewards will come to those who can piece all this information together; the computer scientists have the tools, ability and know-how to strive for and finally reach this end.

Model for the stock market

Let us now set up one of the most accepted models of the stock market. Statistically in the United States about one percent of the population owns about 80 percent of all stock shares, 100 percent of tax exempt government bonds and about 90 percent of corporate bonds. Another 9 percent of the population owns nearly all the remainder, so about 90 percent of the population owns virtually nothing in the way of investment assets. The top one percent makes up the country's plutocracy, whereas the next 9 percent is made up of businessmen. The wealth of the plutocracy comes almost entirely by inheritance, whereas the wealth of the businessmen is derived from such activities as managing corporations, starting businesses, and working in the professions. Because the lower 90 percent of the population own no stocks or bonds to speak about, the stock market may be regarded as a game with players from the plutocracy (one percent of the population with an 80 percent stock ownership) and from the businessmen (nine percent of the population with a 20 percent stock ownership, all more or less). The plutocracy are generally referred to as the "bankers" or "big people" in the market and the businessmen the "public" or "customers." For example, a big person thinks in terms of thousands of shares whereas the public thinks in terms of hundreds of shares.

The following story is well known. Once an out-of-town visitor was being shown the wonders of the New York financial district. When the party arrived at the Battery, one of his guides indicated some handsome ships riding at anchor. He said,

"Look, those are the bankers' and brokers' yachts."

"Where are the customers' yachts?" asked the naïve visitor.

Investment and speculation are said to be two different things, and the prudent man is advised to engage in the one and avoid the other. He perceives that they are different, but they don't seem quite different enough to clear up his problems.

Investment and speculation have been so often defined that a couple more faulty definitions should do no harm, the science of economics having reached a point where further confusion is impossible. Thus,

Speculation is an effort, probably unsuccessful, to turn a little money into a lot.

Investment is an effort, which should be successful, to prevent a lot of money from becoming a little.

If you take a thousand dollars down to Wall Street and attempt to run it up to $25,000 in the course of a year, you are speculating. If you take $25,000 down there and attempt to earn a thousand dollars a year with it you are investing.

There seems to be no way to determine authoritatively the question of how much of a speculator's activity is sheer guesswork, disguised, and how much is sensible. What we are discussing are speculators whose actions are prompted by tape reading, chart reading, statistical analysis, inside information, trading instinct, and all of that.

Let us consider the notion that all of it, or nearly all of it, is actually guesswork. But this cannot be so, it is objected, because a certain few men, in some years, who are speculators, not crooks, make a good thing of it. There are not many of them, but there are, and always have been, a few. And they win. Doesn't this prove that successful speculation is something more than good luck? The answer may be no. And this is why.

There is a mathematical demonstration of what would happen, what *must* happen, if a large number of men were set to playing a game of pure chance against each other. The demonstration is interesting, but the reader must determine for himself whether or not it is analogous to Wall Street speculation. Here it is:

Let us have 400,000 men (and women) engage in this contest at one time. (Something like the number in this country who try being speculators.) We line them up, facing each other in pairs, across a refectory table miles long. Each player is going to play the person facing him a series of games, the game chosen being a matter of pure luck, say matching coins. Two hundred thousand on one side of the table face 200,000 on the other side.

The referee gives a signal for the first game and 400,000 coins flash in the sun as they are tossed. The scorers make their tabulations, and discover that 200,000 people are winners and 200,000 are losers. Then the second game is played. Of the original 200,000 winners, about half of them win again. We now have about 100,000 who have won two games and an equal number who have been so unfortunate as to lose both games. The rest have so far broken even. The simplest thing from now on is to keep our eyes on the winners. (No one is ever much interested in the losers, anyway.)

The third game is played, and of the 100,000 who have won both games half of them are again successful. These 50,000, in the fourth game, are reduced to 25,000, and in the fifth to 12,500. These 12,500 have now won five straight without a loss and are no doubt beginning to fancy themselves as coin flippers. They feel they have an "instinct" for it. However, in the sixth game, 6250 of them are disappointed and amazed to find that they have finally lost. But the victorious 6250 play on and are successively reduced in number until less than a thousand are left. This little band has won some nine straight without a loss, and by this time most of them have at least a local reputation for their ability. People come from some distance to consult them about their method of calling heads and tails, and they modestly give explanations of how they have achieved their success. Eventually there are about a dozen men who have won every single time for about fifteen games. These are regarded as the experts, the greatest coin flippers in history, the men who never lose, and they have their biographies written.

Admittedly, it is preposterous to suggest that stock speculation is like coin flipping. Of course, there is more skill to stock speculation. What is difficult to determine is—how much more?

The stock market in many important ways is analogous to a game of chance in which the players play against each other, and in which the house charges an amount proportional to the total amount played. The stockbroker corresponds to the ''house'' in the game of chance. The commissions that the trader pays to the broker correspond to the percentage of the stakes paid by the player to the house. The speculators (or traders or investors) in the stock market correspond to the players in the game of chance. For the purposes of our model we have divided the speculators into two categories: the big people and the public.

Because the price level of stocks is fluctuating, profits and losses of traders do not balance each other like those of a group playing a game of chance. In a long period, however, the change in the level of all stocks (measured in dollars of constant buying power, so as to remove the effects of inflation) is not large. Generally we may assume that the price level of all stocks is slowly rising over long periods, and this factor together with dividends tends to offset the brokerage commissions, so it can be assumed that the speculators as a whole are playing a game with the odds even. Hence over the long period the people in the market as a group show no profit or loss. Hence what one person gains other people have lost.

In a game of pure change the rules of the game constitute the only pertinent knowledge, so each player has the same chances as any other player. In a game that involves skill as well as chance, the players are not on an equal footing with each other. Of course, it is the relative skill of the players that matter. The stock market corresponds to a game of the latter type, that is, one where skill and judgment are contributing elements as well as chance.

In general common stocks move up or down together; this movement is called the major trend of the market and is measured by averages such as the Dow Jones Industrial Average. If a speculator diversifies his holdings he can eliminate to a considerable degree all considerations except those affecting the market as a whole.

The market as a whole to some extent can be controlled by the big people, and to some extent is beyond the control of any group or governments. However, the big people have a tremendous influence on government monetary and fiscal policies through their lobbying effect on the elected representatives, and these policies definitely affect the major trend. For example, whenever three successive increases are made in any one of the three rates set by the Federal Reserve Board (the rediscount rate, reserve requirements, and stock margin requirements) for some time thereafter the market is likely to suffer a substantial decline.

Superimposed on the major market trends are the trends of the individual stocks. It is in this area that the public is at a distinct disadvantage, for the big people can manipulate the prices of individual stock issues to their advantage.

Thus the big people form a privileged class of speculators, for whom the odds of the game are much more favorable than they are for the others. For example, in the prospectus of each new issue of stock, the Securities and Exchange Commission cause the following statement to be printed to warn the public of one of the legal types of manipulation:

"In connection with this offering, the underwriters may over-allot or effect transactions which stabilize or maintain the market price of the common stock of the company at a level above that which might otherwise prevail in the open market. Such stabilizing, if commenced may be discontinued at any time."

Thus on all new issues the big people (the investment bankers) are legally allowed to manipulate the price to their advantage at the expense of the general public. In addition to the legal forms of manipulation, many other forms are practiced, and when an individual stock moves against the general market, there is a good reason and this reason is known by or due to the big people at the expense of the public. Of course all methods of manipulation that depend upon circulating false information or rumors are dishonest, but there are many other forms that do not depend on such crude tactics and are perfectly legal. Hence speculating in stocks by an individual, no matter how competent he may be in the legitimate aspects of the market, is exceedingly risky, for he is pitting himself against the big people who have access to facts that give them a pronounced and decisive advantage over him.

So much has been written and argued about manipulation of stocks that it is difficult to add much more. The business is based on the fairly sound hypothesis that the public is chiefly interested in buying stocks that are "going up." Thus the manipulators select a stock that they think is underpriced and that has a good story for a "tip" to go with it, and they try to see to it that it "goes up." They also spread the tip, of the truth of which they have carefully convinced themselves, and which may indeed turn out to be true.

If the manipulators make the price rise by washed sales (which are not true transactions and don't cost anything but commissions) then this is a fraud. It is equally a fraud if they spread false information to go with the washed sales. However it is all right to make the price go up by actually buying the stock and paying money for it.

At the conclusion of this first part of the operation they find themselves the owners of a great deal of stock, purchased at ascending prices. At this point the gullible public is supposed to come galloping in to buy the stock from the manipulators at still higher prices. But in a few cases the gullible public acts like an overfed trout and just pays no attention. When this happens the operators, who in the beginning fancied themselves as clever manipulators, wake up one morning to find that they have become involuntary investors.

Manipulation, like other frowned-on practices is not necessarily an easy road to fortune. In fact some manipulations are unsuccessful. A "pool manager," having been supplied with large funds by a "pool" of a dozen men to hoist a certain stock, was having no success whatever. He had bought plenty of stock and the stock was still down. He wrote a letter to each of the members of the pool, explaining at length the hard luck he had run into and asking them each for an additional contribution of five hundred thousand dollars. With this, he assured them, the chestnuts could be pulled out of the fire and a handsome profit would be substituted for an apparent loss. One of his replies read as follows:

Dear Mr. ———

Enclosed please find the check for Five Hundred Thousand Dollars ($500,000) which you requested in yours of the 15th. It was not really necessary for you to assume an apologetic tone. I am sure that you have done your skillful best in this matter, and I am sufficiently experienced to understand that you have encountered reverses which could not be foreseen. Trusting that our enterprise will turn out in the profitable way that you outline, I remain,

Sincerely,

——————— ———————

P.S. That is what I would have written, you (expletive deleted! !) if I had been sucker enough to enclose any check for $500,000.

In short, stock manipulation involves *two phases:* first the *acquisition* by the insiders of the stock of a company at a low price and then the *distribution* to the public of the same stock at a high price. The acquisition phase is actually done quietly over time. However, the distribution phase sometimes is quite spectacular. Witness the following description in the morning newspapers of July 24, 1981.

Takeover issues highlight mixed session on Wall St.

Associated Press

NEW YORK – Takeover targets provided some of the best gains as the stock market halted a three-day slide with a mixed showing yesterday.

The standout stock of the session was Colt Industries, which jumped 21⅛ to 84¼. Penn Central has agreed to acquire the company for $100 a share. Penn Central, like Colt trading for the first time since Monday, dropped 2½ to 43⅛. Colt traded as low as 43¾ this year.

Plight of the small investor

The stock market reflects an amalgam of economic, monetary, and psychological factors. Government policies, wars, political upheavals, inventions, discovery, international credit systems, financial stress and strain, taxation, inventories, equipment, together with the great shifts and variations in mass psychology, hopes and fears, greed, ambition, make up and influence the action of the market. In this sense the stock market represents a most intricate and subtle game comprising weighted subjective judgments and constantly changing patterns and rules. Despite the codified folklore to the contrary and despite its certified respectability, the stock market is primarily a vehicle for the gambler, who is always euphemistically referred to as a speculator, trader, or investor.

In order to support this thesis one only has to read the various investment services urging people to "invest" their money in the stock market. The language and intent of these services appeal to and encourage the gambling instincts of the public. Let us investigate some typical advice given by such advisory services.

A responsible service will always first state that one should never invest in stocks unless he is willing and able to share in the fortunes, for better or worse, of the enterprises selected. In other words, one should be prepared to lose his money. Next the responsible service will state that serious investing should be done with only surplus money,—that is, money left over after one has taken care of his basic needs including an emergency backlog of cash in a savings bank or U.S. savings bonds. In other words, since the chances are high that one is going to lose what he invests in the stock market, he should only invest what he can lose. Finally, the responsible service will say that whatever stocks one buys, he is in for an exciting time in that he will continually be turning to the financial section of the newspaper to see what his stocks are doing. In other words, the stock market provides the enjoyments of gambling in return for the money it takes away from the investor.

Most, if not all, Wall Street publications admit that nothing is more difficult than consistently profiting in the stock market, although they always point out various individuals, partnerships, and closed corporations that have scored great successes. Upon analysis, these lucky winners are nearly always members of the plutocracy,—the "insiders," that upper one percent of society that have organized the financial market over the years to operate in their favor. The remainder of the people are constantly being induced to invest their creatively earned money in the stock market for quick and easy profit, or for a higher income, or as a haven of safety. These people come under the general heading of the "public" or the "small investor." Explanations must constantly be given why the small investor over time nearly always loses, because his losses are apparent. One explanation usually given is that the small investor gives the market less thought than his other activities, and that he is usually careless in his consultations and dealings. Actually the small investor is sold something, usually by an appeal to his gambling instincts. Moreover he is always the victim of sharp practices because the market is controlled as much as it can be controlled by the insiders. However because the market represents nearly the sum total of all human activity and mass psychology, this control cannot be complete but can only be partial. Thus the small investor has a chance of winning; however this probability is less than that probability required to make trading on the stock market a fair game for him. Likewise the probability of the insider winning makes the stock market an advantageous game for him. When a small investor does make a profit he usually assumes that the result was secured through his knowledge and ability instead of realizing that it was most likely the result of chance, that is, luck.

The stock market is a game of chance, with a definite disadvantage to the small investor. These is no such thing as a final answer to security values for the small investor; if he asks a dozen experts he will receive 12 different answers. It often happens that a few days later each expert will alter his conclusions as soon as he is given a chance to reconsider because of a changed condition. The insiders face uncertainty but to a lesser extent; they know beforehand the forthcoming changes in the balance sheets and income statements. The insiders however do face the risks associated with weather, acts of God, and mass psychology. Their total uncertainty, however, is less than that of the small investors. The insiders have a decisive edge in the game of Wall Street.

The only justification for the actions of the small investor is that he is a gambler. If he buys something, or more accurately lets a broker sell him something, and if he makes a profit, then he wants more. If he loses, he most likely will try again. He continues in this way, not knowing that the odds are rigged against him. In fact there is no way of his eliminating his hazards. Of course the more one learns about the risks, the more chance he has to preserve something. Only the insiders will amass fortunes. On the other hand the public by and large has no realization of the dangers, and the result is usually fatal as far as their money is concerned. And it must be this way; certainly the public would not be encouraged to invest in the "rich man's dub" unless the public was paying the bill.

The great force working to the advantage of the insider is the inflationary policy of the U.S. government. Inflation represents the greatest threat to the public for the successful preservation of capital. Of course there are many other threats, such as war, taxation, political changes and revolutions, but these have less immediate threat to the average person than the government-sponsored inflation. The public knows that money placed in a savings bank and allowed to accumulate at compound interest will lose purchasing value. By placing this money in common stocks however it is hoped that the spending power of the money is preserved by the increased dollar value of the capital assets of the company. But it is illusory to assume that the substantial rise in the dollar value of the capital assets represents an increase in their real value. In the 1930's the dollar would buy twenty-five times more than it does in the 1980's. Hence a $10,000 investment made in 1935 that is worth $250,000 in 1982 has not appreciated at all in terms of real value.

Because the public has no confidence in the future value of the dollar, they seek safety in Wall Street. They wish to store for future use today's excess spending power, in such a way that it can be reconverted into usable funds without an overall loss. Unfortunately there is no permanent investment medium available to the public. U.S. Savings Banks pay about 6 percent while the rate of inflation is about 14 percent, so the small depositor (old people, widows, etc.) is guaranteed a 8 percent loss. However this 8 percent loss is certainly better than a total loss, which is almost certainly his ultimate fate on Wall Street. Fortunately most people pull out of Wall Street before they lose all their money, so that something is usually salvaged.

In the broad sense, there is no such thing as a safe investment. The wealth of the world does not increase fast enough with respect to the increase of population to allow payment of compound interest or pyramiding of profits on existing invested capital. For example, if the rich Medici family in Italy six hundred years ago had set aside an investment fund at 5 percent compound interest its value would now be $1,000,000,000,000,000,000, which is many millions of times the existing monetary gold stock of the world today. Clearly such a result is impossible, and adjustments are continually being made through bankruptcies, scaling down of obligations, currency depreciation, and wars to prevent the build-up of invested money at compound interest. The stock market is one of the most efficient means of accomplishing this end.

The risk of inflation is real; however, it is undoubtedly less than the risks of market speculation. A small investor is the participant in an unfair game—a game in which the odds are against him. This fact, however, will not keep the public out of the market, because the gambling instinct is too great.

APPENDIX

Review exercises

1. Define *statistics*.

2. There is a "fundamental" question which we are often trying to answer in statistics: What is it?

3. Define a *parameter*. Define a *statistic*.

4. How do you transform the numbers in a frequency distribution into a relative frequency distribution?

5. What is the *mode* of a set of numbers? What is the *mean* of a set of numbers? What is the *median* of a set of numbers?

6. Define the *variance* of a population. What is the *standard deviation*?

7. Define *outcome*. Define *sample space*. Define *event*. Give examples.

8. If you toss a coin 17 times, how many different *outcomes* are possible? If you toss it *n* times?

9. If a "fair" coin is tossed ten times, among the possible outcomes are *HHHHHHHHHH, HTHTTHHHTT, HHHHHTTTTT,* and *HTHTHTHTHT.* Which is the most likely? The least likely?

10. What is the rule for estimating the variance of a population, on the basis of the numbers in a sample.

11. For a continuous probability distribution, what is the rule for computing the probability of an observation between *a* and *b*?

12. What is the area under a probability density function?

13. What is the standard score or *z* score?

14. How many numbers do you need to completely determine or specify a normal distribution?

15. What is the sampling distribution of the mean, for a given sample size, *n*? How could you approximate it experimentally?

16. What is a null hypothesis? What is a Type One Error? What is a Type Two Error?

17. What is a decision rule? Define *rejection region*. Define *significance level*.

18. For any decision rule, there is a certain chance of a Type One Error and a certain chance of a Type Two Error. If I change to a new decision rule which *reduces* my chance of a Type One Error, what will probably happen to the chance of a Type Two Error?

19. Out of 1,000 tosses of a balanced die, how many 4's would you expect?
    (a) 100   (b) 167   (c) 250   (d) 356

20. If $P(B|A) = 0.8$ and $P(B) = 0.7$, then A and B are:
    (a) mutually exclusive   (b) independent   (c) dependent   (d) all of the above

21. The value for $z = -1.79$ given in the table of areas under the standard normal curve is

(a) 0.4756   (b) 0.4633   (c) 0.4545   (d) 0.4554

22. If the grades on an exam (given in whole numbers) are normally distributed with a mean of 74 and with a standard deviation of 6, the probability that a student picked at random from that class will have a grade between 70 and 79 is about

(a) 26%   (b) 59%   (c) 61%   (d) 73%

23. With reference to Problem 22, the probability that a student picked at random will have a grade between 80 and 84 is about

(a) 14%   (b) 18%   (c) 22%   (d) 26%

24. With reference to Problem 22, the probability that a student picked at random will have a grade greater than 89 is about

(a) 5%   (b) 0.5%   (c) 0.05%   (d) 0.005%

25. With reference to Problem 22, the probability that a student picked at random will have a grade between 65 and 69 is about

(a) 13%   (b) 14%   (c) 15%   (d) 17%

26. How many defective light bulbs would you expect to find in a shipment of 10,000 light bulbs, taken at random from a total output of 100,000, if it is known that 5,500 of the 100,000 bulbs produced were defective?

(a) 55   (b) 100   (c) 550   (d) 1,000

27. What is the probability of getting three aces from a deck of cards by picking three cards, one at a time and without replacement?

(a) $\dfrac{1}{5,675}$ (b) $\dfrac{1}{17,575}$

28. What is the probability of getting fewer than three heads in tossing a balanced coin three times?

(a) $\dfrac{1}{8}$ (b) $\dfrac{2}{8}$ (c) $\dfrac{3}{8}$ (d) $\dfrac{4}{8}$

29. What is the probability of getting three aces by picking three cards, one at a time from a deck of cards, if each card picked is always replaced and the cards are reshuffled before the next card is picked?

(a) $\dfrac{3}{52}$ (b) $\dfrac{1}{52}$ (c) $\dfrac{1}{5,204}$ (d) $\dfrac{1}{2,197}$

30. The mean strength of a sample of 100 pieces of rope produced by a company is found to be 98.5 lbs. with a standard deviation of 8 lbs. The hypothesis that $\mu = 100$ lbs. at the 5% level of significance should be

(a) accepted   (b) rejected

31. Would the hypothesis in Problem 30 be accepted at the 1% level of significance?

(a) yes   (b) no

32. With reference to Problem 30, what is the smallest value and the largest value for $\bar{x}$ consistent with H_0?

(a) 96.2 lbs. and 103.8 lbs., respectively
(b) 97.3 lbs. and 102.3 lbs., respectively
(c) 98.4 lbs. and 101.6 lbs., respectively
(d) 98.6 lbs. and 101.4 lbs., respectively

33. An urn contains marbles which are either red or blue. To test the hypothesis of equal proportions of these colors, we agree to sample 64 marbles with replacement, noting the colors drawn and adopt the following decision rule: (*1*) accept the hypothesis if between 28 and 36 red marbles are drawn; (*2*) reject the hypothesis otherwise.

(a) Find the probability of rejecting the hypothesis when it is actually correct.

(b) Interpret graphically the decision rule and the result obtained in (a).
Ans. (*a*) .2606

34. 64 heads are obtained in 100 tosses of a coin. Is this coin fair? (Make a decision at the 1% and at the 5% level of significance.)
(a) yes   (b) no

35. Given the empirical frequency distribution

Amount	Frequency
0-$25	7
$25-$50	20
$50-$75	43
$75-$100	30

then the average is:
(a) $37.50   (b) $61.50   (c) $62.50   (d) $87.50;
the median is:
(a) $49.50   (b) $61.50   (c) $62.50   (d) $62.87;
the mode is:
(a) $59.50   (b) $62.50   (c) $65.47   (d) either b or c

36. In his long short story *The End of the Tether,* Joseph Conrad created the character of George Massy, a ship's engineer, who won 'the second great prize in the Manilla lottery', bought his own ship with the proceeds and became obsessed with the idea of winning again:

'With his elbows propped, his head between his hands, he seemed to lose himself in the study of an abstruse problem in mathematics. It was the list of winning numbers from the last drawing of the great lottery which had been the one inspiring fact of so many years of his existence . . . There was in them, as in the experience of life, the fascination of hope, the excitement of a half-penetrated mystery, the longing of a half-satisfied desire.

'For days together on a trip, he would shut himself up in his berth with them . . . and he would weary his brain poring over the rows of disconnected figures, bewildering by their senseless sequence, resembling the hazards of destiny itself. He nourished a conviction that there must be some logic lurking somewhere in the results of chance. He thought he had seen its very form . . .

Nine, nine, nought, four, two. He made a note. The next winning number of the great prize was forty-seven thousand and five. These numbers of course would have to be avoided in the future when writing to Manilla for the tickets. He mumbled, pencil in hand . . . "and five. Hm . . . Hm." He wetted his finger: the papers rustled. Ha! But what's this? Three years ago, in the September drawing, it was the number nine, nought, four, two that took the first prize. Most remarkable. There was a hint there of a definite rule! He was afraid of missing some recondite principle in the overwhelming wealth of his material. What could it be? and for half an hour he would remain dead still, bent low over his desk, without twitching a muscle.'

Was Massy right in principle? By studying the numbers involved, can we accumulate huge sums by gambling?

37. Comment on the following system in roulette. It was devised by William Jaggers, a British engineer, and it earned him a profit of 1,500,000 francs at the end of the nineteenth century. He hired six men to jot down the winning numbers for a whole month and when he went through them, he knew immediately that he had found a way of beating the house advantage. For some of them were coming up much more frequently than the Law of Probability said they should. He concentrated his bets on these and won consistently. For, as he had suspected, the tables were biased, not because the owners wished to cheat, but because they had allowed the balance to get out of true. The knowledge gave Mr. Jaggers an unbeatable advantage.

APPENDIX

One hundred review problems

1. The arithmetic mean of the data 4, 5, 6, 7 is
   (a) 5   (c) 5.5
   (b) 6   (d) 5.25
2. The standard deviation of the data 4, 5, 6 is
   (a) 1   (c) 0.816
   (b) 2/3   (d) 0
3. The standard deviation of the data 100, 100, 100, 100, 100, 100, 100, 100, 100, 100 is
   (a) infinite   (c) 0
   (b) 10   (d) $\sqrt{10}$
4. For major types of popular publications, the classic paperbacks, soft covers, magazines, issues of *Time,* issues of *Newsweek,* the Bible, newspapers, are
   (a) mutually exclusive but not exhaustive
   (b) exhaustive but not mutually exclusive
   (c) mutually exclusive and exhaustive
   (d) neither mutually exclusive nor exhaustive
5. For age at death of Americans, the class intervals 0-1, 2-5, 6-10, 11-20, 21-40, 41-60, 61-80 are
   (a) exhaustive   (c) equal
   (b) geometric   (d) not exhaustive
6. For the data 1, 1, 1, 1, 2, 2, 2, 2, 2, 3, 3, 3, 3, 3, 4 the relative frequency of the score 2 is
   (a) 2/15   (c) 5/14
   (b) 1/4   (d) 1/3
7. For the data 2, -3, 1 the value $\Sigma(x - \bar{x})^2$ is
   (a) 15   (c) -4
   (b) 0   (d) 14
8. A teacher obtained scores on an achievement test for 4 students as follows: 110, 110, 110, 110.
   For these scores what are:
   (a) the mean?
   (b) variance?
   (c) standard deviation?
   (d) z scores for each?

9. The measure of central tendency which reflects the influence of extreme scores is the
  (a) mean
  (b) median
  (c) mode
  (d) midpoint

10. For a sample size 100 let f' denote relative frequency. Then
  (a) $\Sigma f' = 100\bar{x}$ (c) $\Sigma f' = 100$
  (b) $\Sigma f' = 1$ (d) $\Sigma f' = \bar{x}/100$

11. A teacher tested the creativity of children in her class. The creativity scores were:

    9, 6, 5, 7, 3.

Compute the mean, variance and standard deviation of these creativity scores.

12. A given set of numerical data in the form of a frequency distribution has only one value of n but many x and _____ values. Fill in the blank.
  (a) s (c) f
  (b) $\bar{x}$ (d) s^2

13. Essay and multiple choice exam scores were available for 4 students. Convert these to z scores and find a combined score for each student.

	Essay	MC
1.	6	3
2.	3	5
3.	2	1
4.	5	3

14. In administering three exams, a four week, eight week and twelve week, an instructor wanted to give increasing weight to the importance of test. The eight week was twice as important as the four week, the twelve week three times as important as the four week, exam. One individual scored in percentages 90 on the first exam, 81 on the second and 70 on the third. A second student scored 26 on the first exam, 68 on the second and 100 on the third. Which student did better in the course and why? (show your computation)

15. In English the mean was 73 and the standard deviation 12 whereas in History the mean was 58 and the standard deviation was 4. If John's score was 80 in English and 61 in History, then John did
  (a) better in English than History
  (b) better in History than English
  (c) the same in History and English
  (d) worse in History than English

16. If the mean is 75 and the variance is 9 then the score 70 can be converted into the standard score
  (a) $-5/3$ (c) $-5/9$
  (b) $-5/8$ (d) $5/3$

17. Refer to the grouped frequency distribution of achievement test scores to the right. From this distribution one could say:

class interval	f
95-99	3
90-94	6
85-89	12
80-84	24
75-79	12
70-74	6
65-69	3

(a) 21 individuals obtained a score of 77 or below.

(b) 12 individuals obtained a score of 82.

(c) 9 individuals obtained a score of 90 or above.

(d) 0 individuals scored 71.

18. The _____ separates the upper 37% in a distribution from the lower 73% in a distribution.

(a) 72 percentile

(b) 73 decile

(c) 74 percentile

(d) none of the above

19. In a histogram drawn with the width of each bar equal to 1:

(a) the height of each bar represents the frequency of that class.

(b) lines are drawn connecting the midpoints of successive classes.

(c) the sum of the areas of all bars is equal to 1.

(d) the frequency of extreme classes is always less than that of the middle classes.

20. For the data 1, 3, 3, 5, 4, 2, 4, 3, 2, the histogram is

(a) symmetrical   (c) equally likely

(b) skewed   (d) bimodal

21. Four aces are placed in a hat: hearts, spades, diamonds, and clubs. One is drawn at random, the result is recorded, and the card is replaced. Then a second draw is made, and the result is recorded. Draw a tree to represent this experiment. How many individual outcomes are there? Which outcomes are included in the event "At least one heart"? Which outcomes are included in the event "Both cards are red"? What are the probabilities of these two events?

22. Given the grades and distribution of scores for a class in mathematics, compute the 9th decile, 3rd quartile, and 80th percentile.

Grades	Frequency of Students
D = 81-85	5
C = 86-90	10
B = 91-95	4
A = 96-100	1

	9th decile	3rd quartile	80th percentile
(a)	81.75	86.25	95.25
(b)	81.00	86.00	90.00
(c)	95.50	85.50	90.50
(d)	94.25	90.50	91.75
(e)	none of the above		

23. Scores of nine students on a test are:

$$\begin{array}{ccc} 30 & 45 & 80 \\ 90 & 35 & 20 \\ 16 & 85 & 15 \end{array}$$

The median score on this test is:

(a) 20.0
(b) 51.3
(c) 35.0
(d) 45.0
(e) 80.0

24. Given the data

$$\begin{array}{ccc} 167 & 194 & 219 \\ 168 & 195 & 227 \\ 171 & 200 & 232 \\ 179 & 204 & 247 \\ 186 & 211 & 260 \end{array}$$

The range is: The standard deviation is:

(a) $167   (a) 28.63
(b) $46.50   (b) 27.66
(c) $93   (c) 764.80
(d) $260   (d) 11,472

25. (a) Toss 4 coins fifty times and tabulate the number of heads at each toss. (b) Construct a frequency distribution showing the number of tosses in which 0, 1, 2, 3, 4 heads appeared. (c) Construct a percentage distribution corresponding to (b). (d) Compare the percentage obtained in (c) with the theoretical ones 6.25%, 25%, 37.5%, 25%, 6.25% (proportional to 1, 4, 6, 4, 1) arrived at by rules of probability. (e) Construct graphical presentation of the distributions in (b) and (c).

26. Three teachers of economics reported mean examination grades of 79, 74 and 82 in their classes which consisted of 32, 25 and 17 students respectively. Determine the mean grade for all the classes.
Ans. 78

27. A student's grades in the laboratory, lecture and recitation parts of a physics course were 71, 78 and 89 respectively. (a) If the weights accorded these grades are 2, 4 and 5 respectively, what is an appropriate average grade? (b) What is the average grade if equal weights are used?   *Ans.* (a) 82, (b) 79

28. A set of numbers consists of six 6's, seven 7's, eight 8's, nine 9's and ten 10's. What is the arithmetic mean of the numbers?   *Ans.* 8.25.

29. (a) By adding 5 to each of the numbers in the set 3, 6, 2, 1, 7, 5 we obtain the set 8, 11, 7, 6, 12, 10. Show that the two sets have the same standard deviations but different means. How are the means related?

(b) By multiplying each of the numbers 3, 6, 2, 1, 7, 5 by 2 and then adding 5, we obtain the set 11, 17, 9, 7, 19, 15. What is the relationship between the standard deviations and the means for the two sets?

    (c) What properties of the mean and standard deviation are illustrated by the particular sets of numbers in (a) and (b)?

30. Find the standard deviation of the set of numbers in the arithmetic progression 4, 10, 16, 22, . . . , 154.

31. Suppose the probability of hitting the "bull's-eye" on a single shot is .4. What is the smallest number of shots which must be fired in order to have a probability of at least .9 of hitting the "bull's-eye" at least once?

32. Three urns contain respectively one white, and two black balls; three white and one black ball; two white and three black balls. One ball is taken from each urn. What is the probability that among the balls drawn there are two white and one black?

33. A fair coin is tossed until for the first time the same result appears twice in succession. Find the probability of the following events:

    (a) The experiment ends before the sixth toss

    (b) An even number of tosses is required

34. A retail hardware dealer kept count of the defective one-inch wood screws purchased from a certain manufacturer and found that one in fifty screws are defective. What is the probability that in a random sample of 2500 there will be 64 or more that are defective?

35. A man wishes to unlock his door in the dark, and has a ring of m keys. The correct key can only be found by trying it in the lock. He can select successive keys at random, or by rotation key by key around the ring. Find the probability of success at the nth trial for each method. Which method is thus more efficient?

36. An urn contains 12 red balls, 10 white balls, 15 blue balls, and 3 black balls. If one ball is drawn at random (each ball has a probability of 1/40), find the probability that the ball is

    (a) red, white, or blue

    (b) neither white nor black

    (c) black or blue

37. An interviewer working for a public opinion poll randomly selects two persons in succession from a group of 30 including 18 favoring Carter, 8 favoring Ford, and 4 who are undecided.

    (a) what is the probability that both favor Carter?

    (b) What is the probability that both favor Ford?

    (c) What is the probability that one favors Carter and that one favors Ford?

    (d) What is the probability that one favors Carter and the other is undecided?

38. A coin is flipped 674 times and shows up heads 536 times. What's the probability of a head on the 537th flip?

39. Two dice are rolled. One shows up 5. What's the probability the other will show up even?

40. What is the probability of getting more than 3 in rolling two balanced dice simultaneously?

(a) $\dfrac{4}{36}$ (c) $\dfrac{1}{4}$

(b) $\dfrac{1}{9}$ (d) $\dfrac{11}{12}$

41. A ball is drawn at random from a box containing 10 red, 30 white, 20 blue and 15 orange marbles. Find the probability that it is (a) orange or red, (b) not red or blue, (c) not blue, (d) white, (e) red, white or blue.   *Ans.* (a) 1/3, (b) 3/5, (c) 11/15, (d) 2/5, (e) 4/5

42. Two marbles are drawn in succession from the box of the preceding problem, replacement being made after each drawing. Find the probability that (a) both are white, (b) the first is red and the second is white, (c) neither is orange, (d) they are either red or white or both (red and white), (e) the second is not blue, (f) the first is orange, (g) at least one is blue, (h) at most one is red, (i) the first is white but the second is not, (j) only one is red.
Ans. (a) 4/25, (b) 4/75, (c) 16/25, (d) 64/225, (e) 11/15, (f) 1/5, (g) 104/225, (h) 221/225, (i) 6/25, (j) 52/225

43. Work the preceding problem if there is no replacement after each drawing.
Ans. (a) 29/185, (b) 2/37, (c) 118/185, (d) 52/185, (e) 11/15, (f) 1/5, (g) 86/185, (h) 182/185, (i) 9/37, (j) 26/111

44. There are 40 marbles in an urn: 18 blue, 10 red, and 12 white. The probabilities of getting a red or a white; a red or a blue; a blue or a white marble, on any draw are:

	red or white	red or blue	blue or white
(a)	3/10	9/20	3/10
(b)	11/20	15/20	7/10
(c)	1	1	1
(d)	11/20	7/10	15/20
(e)	none of the above		

45. Two dice are rolled once. What is the probability of getting an even number on either die on one roll of the two dice?

    (a) 1/6

    (b) 1/2

    (c) 1/3

    (d) 1/4

    (e) none of the above

46. Three dice are thrown successively. All three are unbiased dice. Which of the following statements is true?

    (a) The probability of three ⚀ is the same as the probability of ⚀ on the first throw, ⚁ on the second throw, and ⚂ on the third throw.

    (b) The probability of three ⚂ is 1/216.

    (c) The probability of the first two of the three dice turning up ⚀ is 1/36.

(d) all of the above.

(e) none of the above.

47. Students who misbehave in school may be detained (D), reprimanded (R), or ignored (I). Given $P(D) = .60$, $P(R) = .30$, $P(I) = .10$. No student may receive more than one of the three possible treatments. Which of the following probability statements is true? (Note: A prime indicates the contrary event.)

(a) $P(D \cap R) = .18$

(b) $P(I \cap R) = .03$

(c) $P(D' \cup R) = .30$

(d) $P(D \cup R') = .70$

(e) $P(I \cup R) = .37$

48. Find the probability of scoring a total of 7 points (a) once, (b) at least once, (c) twice in two tosses of a pair of fair dice. *Ans.* (a) 5/18, (b) 11/36, (c) 1/36

49. The odds in favor of A winning a game of chess against B are 3:2. If three games are to be played, what are the odds (a) in favor of A's winning at least two games out of the three, (b) against A losing the first two games to B? *Ans.* (a) 81:44, (b) 21:4

50. A box contains 9 tickets numbered from 1 to 9 inclusive. If 3 tickets are drawn from the box one at a time, find the probability that they are alternately either odd, even, odd or even, odd, even. *Ans.* 5/18

51. A friend once consulted Galileo on the following difficulty: When three dice are thrown the number 9 and the number 10 can each be produced by six different combinations. Nevertheless experience shows that the number 10 is thrown more often than the number 9. Galileo resolved this difficulty. Can you? *Ans.* $P(9) = \dfrac{25}{216}$ $P(10) = \dfrac{27}{216}$

52. What is more probable:

(a) that at least one ace comes up in 4 throws of 1 die?

(b) that at least one ace comes up in 24 throws of 2 dice?

Ans. (a) $1 - \left(\dfrac{5}{6}\right)^4$ (b) $1 - \left(\dfrac{35}{36}\right)^{24}$

53. An "honest" die is tossed n times. Find the expected (average) number of times when a 6 is followed by a number less than or equal to 3.

Ans. $\dfrac{n-1}{12}$

54. Find the probability that the birthday of 12 people will fall within 12 different calendar months. *Ans.* $12!/12^{12}$

55. Find the probability that the birthdays of 6 people will fall within exactly 2 calendar months (any 2).

Ans. $\dfrac{2^6 - 2}{12^6} \binom{12}{2}$

56. Five people are selected at random. What is the probability that at least three of them have the same birthday?

57. Eight people are selected at random. What is the probability that exactly two of them have the same birth month?

58. If it rains, an umbrella salesman can earn $30 per day. If it is fair he can lose $6 per day. What is his expectation if the probability of rain is .3?   *Ans.* $4.80 per day

59. *A* and *B* play a game in which they toss a fair coin three times. The one obtaining heads first wins the game. If *A* tosses the coin first and if the total value of the stakes is $20, how much should be contributed by each in order that the game be considered fair?   *Ans. A,* $12.50; *B,* $7.50

60. Find the expected value of $(x + y)^2$ where x and y are independent standard normal random variables.

61. What is a fair price to pay to enter a game in which one can win $25 with probability .2 and $10 with probability .4?   *Ans.* $9

62. In how many ways can 3 men and 3 women be seated at a round table if (a) no restriction is imposed, (b) two particular women must not sit together, (c) each woman is to be between two men?   *Ans.* (a) 120, (b) 72, (c) 12

63. In how many ways can 7 books be arranged on a shelf if (a) any arrangement is possible, (b) 3 particular books must always stand together, (c) two particular books must occupy the ends? *Ans.* (a) 5040, (b) 720, (c) 240

64. In how many ways can 2 men, 4 women, 3 boys and 3 girls be selected from 6 men, 8 women, 4 boys and 5 girls if (a) no restrictions are imposed, (b) a particular man and woman must be selected? *Ans.* (a) 42,000, (b) 7000

65. How many different committees of 3 men and 4 women can be formed from 8 men and 6 women? *Ans.* 840

66. In how many ways can 6 questions be selected out of 10?   *Ans.* 210

67. How many numbers consisting of five different digits each can be made from the digits 1, 2, 3, . . . , 9 if (a) the numbers must be odd, (b) the first two digits of each number are even? *Ans.* (a) 8400, (b) 2520

68. Find (a) $E(x)$, (b) $E(x^2)$, (c) $E[(x - \bar{x})^2]$, and (d) $E(x^3)$ for the following probability distribution.

x	-10	-20	30
$P(x)$	1/5	3/10	1/2

     Ans. (a) 7, (b) 590, (c) 541, (d) 10,900

69. A random variable assumes the value 1 with probability p, and 0 with probability $q = 1 - p$. Prove that (a) $E(x) = p$, (b) $E[(x - \mu)^2] = pq$.

70. In working with probability density functions, one repeatedly mentions "the area over the interval." Did you mean the area divided by the interval? *Ans.* No! This phrase seems to generate some confusion, and more precisely one could say "the area between the horizontal axis and the curve bounded by the two sides of the interval."

71. What is the probability of getting a 9 exactly once in 3 throws with a pair of dice?   *Ans.* 64/243

72. Find the probability of getting a total of 11 (a) once, (b) twice in two tosses of a pair of fair dice. *Ans.* (a) 17/162, (b) 1/324

73. The *standard normal curve* has three characteristics. These characteristics concern the total area under the curve, the mean of the curve, the standard deviation of the curve. Which one of the choices below is correct?

	Total area	Mean	Standard Deviation
(a)	1.0	1.0	0.0
(b)	1.0	0.0	1.0
(c)	1.0	0.0	0.0
(d)	None of the above		

74. Find the area under the normal curve (a) to the left of $z = -1.78$, (b) to the left of $z = .56$, (c) to the right of $z = -1.45$, (d) corresponding to $z \geq 2.16$, (e) corresponding to $-.80 \leq z \leq 1.53$, (f) to the left of $z = -2.52$ and to the right of $z = 1.83$.   *Ans.* (a) .0375, (b) .7123, (c) .9265, (d) .0154, (e) .7251, (f) .0395

75. If the mean and the standard deviation of a normal distribution are 182 and 12, respectively, an x-value of 214 corresponds to a z-value of
    (a) 0.32   (c) 2.67
    (b) 1.81   (d) 3.09

76. Find the area under the normal curve between (a) $z = -1.20$ and $z = 2.40$, (b) $z = 1.23$ and $z = 1.87$, (c) $z = -2.35$ and $z = -.50$.   *Ans.* (a) .8767, (b) .0786, (c) .2991

77. Suppose that you are taking samples from the standard normal distribution. What is the probability of observing a value between -1.96 and -2.58?

78. Find (a) the mean and (b) the standard deviation on an examination in which grades of 70 and 88 correspond to standard scores of $-.6$ and 1.4 respectively.   *Ans.* (a) 75.4, (b) 9

79. On a statistics examination the mean was 78 and the standard deviation was 10. (a) Determine the standard scores of two students whose grades were 93 and 62 respectively. (b) Determine the grades of two students whose standard scores were $-.6$ and 1.2 respectively.   *Ans.* (a) 1.5, -1.6; (b) 72, 90

80. Define each of the following in a few words:
    (a) Type One Error.
    (b) Type Two Error.
    (c) Sample space.
    (d) Sampling distribution of the mean.

81. The null hypothesis is assumed, and on this assumption the level of significance is determined for a particular obtained result. It can be stated that:
    (a) The null hypothesis is rejected when the result is significant beyond the .05 level.
    (b) A result which is significant at the .02 level is *more* significant than a result significant at the .01 level.

(c) A result which is significant at the .10 level is expected to occur by chance about ten times in 100 experiments when the null hypothesis is correct.

(d) A result which is significant at the .05 level is 5 per cent certain to be significant.

82. The average height of men 50 years ago was 170 cm ($= \mu$) with a standard deviation of 6.7 cm ($= \sigma$). Assume that today the standard deviation is still 6.7 cm. It has been proclaimed that today the average height of men is 172 cm or less. After examining the heights of 50 randomly selected men you found a sample mean of 173.5 ($= \bar{x}$). Does this sample entitle you to deny the statement that μ is 172 or less?

83. A Type One Error will occur when we:

(a) correctly accept the null hypothesis.

(b) correctly reject the null hypothesis.

(c) incorrectly accept the null hypothesis.

(d) incorrectly reject the null hypothesis.

(e) none of the above.

84. On a statistics midterm, the 36 students in section 1 received an average grade of 83, with standard deviation of 8. The 35 students in section 2 received an average grade of 76, with standard deviation of 6. Is there a difference between the two sections at the 5% level of significance?

85. A study of Shakespeare's plays involved looking at each group of 100 consecutive words and counting the number of *nouns* in the group. This index, which we shall call N, was found to be very nearly normally distributed, with a mean of 17 and standard deviation of 6. An unsigned play written in a similar style is found, and a noun count is made for a sample of 50 100-word passages. The average of these 50 measurements is 19.7. Do you think this play was by Shakespeare? Support your conclusion with a statistical test.

86. An objective examination contains 10 questions. Each question has five answers; only one of the five answers is correct.

(a) What is the expected number of correct answers a student without preparation could make? _____

(b) Set up a test of a statistical hypothesis, so designed that the completely unprepared student will fail the examination. (Let $\alpha = .05$).

(1) State H_0

(2) State H_1

(3) What is the critical region?

(c) What is the probability that a student sufficiently prepared to answer 30 percent of the questions correctly, will fail? _____.

If he does fail is it an example of a Type I or Type II Error.

87. On an examination given to students at a large number of different schools, the mean grade was 74.5 and the standard deviation was 8.0. At one

particular school where 200 students took the examination, the mean grade was 75.9. Discuss the significance of this result at a .05 level from the viewpoint of (a) a one-tailed test and (b) a two-tailed test, explaining carefully your conclusions on the basis of these tests.   *Ans.* The result is significant at a .05 level in both a one-tailed and two-tailed test.

Repeat the above if the significance level is .01.
Ans. The result is significant at a .01 level in a one-tailed test but not in a two-tailed test.

88. A pair of dice is tossed 100 times and it is observed that "sevens" appear 23 times. Test the hypothesis that the dice are fair, i.e. not loaded, using (a) a two-tailed test and (b) a one-tailed test and a significance level of .05. Discuss your reasons, if any, for preferring one of these tests over the other.
Ans. (a) We cannot reject the hypothesis at .05 level. (b) We can reject the hypothesis at .05 level.

89. A manufacturer claimed that at least 95% of the equipment which he supplied to a factory conformed to specifications. An examination of a sample of 200 pieces of equipment revealed that 18 were faulty. Test his claim at a significance level of (a) .01, (b) .05.
Ans. We can reject the claim at both levels using a one-tailed test.

90. The percentage of A's given in a physics course at a certain university over a long period of time was 10%. During one particular term there were 40 A's in a group of 300 students. Test the significance of this result at a level of (a) .05, (b) .01.
Ans. Using a one-tailed test, the result is significant at a .05 level but is not significant at a .01 level.

91. Out of 800 families with 5 children each, how many would you expect to have (a) 3 boys, (b) 5 girls, (c) either 2 or 3 boys. Assume equal probabilities for boys and girls.   *Ans.* (a) 250, (b) 25, (c) 500

92. A company manufactures rope whose breaking strengths have a mean of 300 lb and standard deviation 24 lb. It is believed that by a newly developed process the mean breaking strength can be increased.

   (a) Design a decision rule for rejecting the old process at a .01 level of significance if it is agreed to test 64 ropes.
   (b) Under the decision rule adopted in (a), what is the probability of accepting the old process when in fact the new process has increased the mean breaking strength to 310 lb? Assume the standard deviation is still 24 lb.

Ans. (a) Reject H_0 if the mean breaking strength of 64 ropes exceeds 307.0 lb. Accept H_0 otherwise.
   (b) β = (area under right-hand normal curve to left of $z = -1.00$) = .1587. This is the probability of accepting H_0: $\mu = 300$ lb when actually H_1: $\mu = 310$ lb is true, i.e. it is the probability of making a Type II Error.

93. Define each of the following as accurately as you can.
    (a) Significance level.   (f) Null hypothesis.
    (b) Sample space.   (g) Outcome.
    (c) Rejection region.   (h) Type One Error.
    (d) Event.   (i) Decision rule.
    (e) Type Two Error.

94. Suppose that Volvo car sales for the past eleven years are as follows (in thousands of cars):

10 years ago	0
9 years ago	1
8 years ago	1
7 years ago	2
6 years ago	1
5 years ago	3
4 years ago	7
3 years ago	16
2 years ago	27
1 year ago	24
This year	33

Would it be surprising if 90 percent of the cars sold in the past eleven years were still on the road? What is the most "representative" age in this group of cars, the age such that half of the cars are newer than it and half of the cars are older? This is called the *median* age.

95. Why is a Type One Error sometimes called a *false alarm*? Why is a Type Two Error sometimes called a *miss*?

96. A language student who is a statistician notes that the number of words he can learn in an hour is normally distributed, with a mean of 14 and a standard deviation of $\sqrt{12}$. One Sunday he switches to a health food diet and for the next *three hours* averages 19.3 words per hour. Interpret this number, 19.3, as the mean of a sample of size 3 from a population. Should he conclude that his learning rate has changed?
Hints: What would be the probability distribution for means of samples of size 3, taken from the population, "the number of words learned in 1 hour," for this student before his dietary change? Is it likely that you would see a number as extreme as 19.3 in this derived probability distribution? Compute a standard score to support your argument.

97. Fifteen classes see a new film, "Highway Safety." For the next 24 months, a record is kept of the number of moving traffic violations for each class, and the violation rate is compared with the national average. Of the 15 classes, 11 do better than the national average and 4 do worse. Is this a significant result? Use a 0.05 significance level, and be specific about your logic.

98. Suppose one wanted to run a z test with a 0.08 significance level. One should reject the null hypothesis for any observed z score greater than _____ or less than _____.

99. Suppose that grades in a large course are normally distributed with mean 70 and standard deviation 10. Grades between 90 and 100 get an A, between 80 and 90 get a B, between 70 and 80 get a C, between 60 and 70 get a D, and below 60 the grade is F.

    (a) Find the z score associated with each of the cut-points: 60, 70, 80, 90.

    (b) Find the probability of observing each of the five grades. (Remember that the probability of observing a grade between, say, 60 and 70 is the area under the curve between these two numbers.)

100. What is it that a z score tells you?

Ans. A z score always tells you the location of one particular observation in a known distribution. The z score tells you how many standard deviations away from the mean your observation lies. Sometimes, of course, the observation itself is a sample mean, and the distribution is the sampling distribution of sample means of that size.

GREEK ALPHABET

Letters		Names	Letters		Names
A	α	alpha	N	ν	nu
B	β	beta	Ξ	ξ	xi
Γ	γ	gamma	O	o	omicron
Δ	δ	delta	Π	π	pi
E	ϵ	epsilon	P	ρ	rho
Z	ζ	zeta	Σ	σ	sigma
H	η	eta	T	τ	tau
Θ	θ	theta	Υ	υ	upsilon
I	ι	iota	Φ	ϕ	phi
K	κ	kappa	X	χ	chi
Λ	λ	lambda	Ψ	ψ	psi
M	μ	mu	Ω	ω	omega

TABLE I   BINOMIAL COEFFICIENTS $\binom{n}{r}$

n	r										
	0	1	2	3	4	5	6	7	8	9	10
0	1										
1	1	1									
2	1	2	1								
3	1	3	3	1							
4	1	4	6	4	1						
5	1	5	10	10	5	1					
6	1	6	15	20	15	6	1				
7	1	7	21	35	35	21	7	1			
8	1	8	28	56	70	56	28	8	1		
9	1	9	36	84	126	126	84	36	9	1	
10	1	10	45	120	210	252	210	120	45	10	1
11	1	11	55	165	330	462	462	330	165	55	11
12	1	12	66	220	495	792	924	792	495	220	66
13	1	13	78	286	715	1287	1716	1716	1287	715	286
14	1	14	91	364	1001	2002	3003	3432	3003	2002	1001
15	1	15	105	455	1365	3003	5005	6435	6435	5005	3003

BINOMIAL PROBABILITIES FOR $p = 0.5$

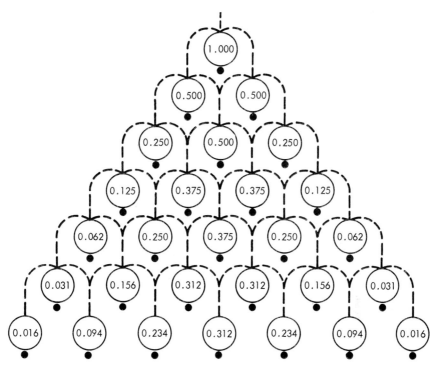

The top row corresponds to $n = 0$, the next row to $n = 1$, the next row to $n = 2$, and so on. The probabilities given in each row (from $n = 2$ on) can be found in the $p = 0.500$ column in the next table.

TABLE II   BINOMIAL PROBABILITIES

n	x	0.050	0.100	0.200	0.300	0.400	0.500	0.600	0.700	0.800	0.900	0.950
2	0	0.902	0.810	0.640	0.490	0.360	0.250	0.160	0.090	0.040	0.010	0.002
	1	0.095	0.180	0.320	0.420	0.480	0.500	0.480	0.420	0.320	0.180	0.095
	2	0.002	0.010	0.040	0.090	0.160	0.250	0.360	0.490	0.640	0.810	0.902
3	0	0.857	0.729	0.512	0.343	0.216	0.125	0.064	0.027	0.008	0.001	
	1	0.135	0.243	0.384	0.441	0.432	0.375	0.288	0.189	0.096	0.027	0.007
	2	0.007	0.027	0.096	0.189	0.288	0.375	0.432	0.441	0.384	0.243	0.135
	3		0.001	0.008	0.027	0.064	0.125	0.216	0.343	0.512	0.729	0.857
4	0	0.815	0.656	0.410	0.240	0.130	0.062	0.026	0.008	0.002		
	1	0.171	0.292	0.410	0.412	0.346	0.250	0.154	0.076	0.026	0.004	
	2	0.014	0.049	0.154	0.265	0.346	0.375	0.346	0.265	0.154	0.049	0.014
	3		0.004	0.026	0.076	0.154	0.250	0.346	0.412	0.410	0.292	0.171
	4			0.002	0.008	0.026	0.062	0.130	0.240	0.410	0.656	0.815
5	0	0.774	0.590	0.328	0.168	0.078	0.031	0.010	0.002			
	1	0.204	0.328	0.410	0.360	0.259	0.156	0.077	0.028	0.006		
	2	0.021	0.073	0.205	0.309	0.346	0.312	0.230	0.132	0.051	0.008	0.001
	3	0.001	0.008	0.051	0.132	0.230	0.312	0.346	0.309	0.205	0.073	0.021
	4			0.006	0.028	0.077	0.156	0.259	0.360	0.410	0.328	0.204
	5				0.002	0.010	0.031	0.078	0.168	0.328	0.590	0.774
6	0	0.735	0.531	0.262	0.118	0.047	0.016	0.004	0.001			
	1	0.232	0.354	0.393	0.303	0.187	0.094	0.037	0.010	0.002		
	2	0.031	0.098	0.246	0.324	0.311	0.234	0.138	0.060	0.015	0.001	
	3	0.002	0.015	0.082	0.185	0.276	0.312	0.276	0.185	0.082	0.015	0.002
	4		0.001	0.015	0.060	0.138	0.234	0.311	0.324	0.246	0.098	0.031
	5			0.002	0.010	0.037	0.094	0.187	0.303	0.393	0.354	0.232
	6				0.001	0.004	0.016	0.047	0.118	0.262	0.531	0.735
7	0	0.698	0.478	0.210	0.082	0.028	0.008	0.002				
	1	0.257	0.372	0.367	0.247	0.131	0.055	0.017	0.004			
	2	0.041	0.124	0.275	0.318	0.261	0.164	0.077	0.025	0.004		
	3	0.004	0.023	0.115	0.227	0.290	0.273	0.194	0.097	0.029	0.003	
	4		0.003	0.029	0.097	0.194	0.273	0.290	0.227	0.115	0.023	0.004
	5			0.004	0.025	0.077	0.164	0.261	0.318	0.275	0.124	0.041
	6				0.004	0.017	0.055	0.131	0.247	0.367	0.372	0.257
	7					0.002	0.008	0.028	0.082	0.210	0.478	0.698
8	0	0.663	0.430	0.168	0.058	0.017	0.004	0.001				
	1	0.279	0.383	0.336	0.198	0.090	0.031	0.008	0.001			
	2	0.051	0.149	0.294	0.296	0.209	0.109	0.041	0.010	0.001		
	3	0.005	0.033	0.147	0.254	0.279	0.219	0.124	0.047	0.009		
	4		0.005	0.046	0.136	0.232	0.273	0.232	0.136	0.046	0.005	
	5			0.009	0.047	0.124	0.219	0.279	0.254	0.147	0.033	0.005
	6			0.001	0.010	0.041	0.109	0.209	0.296	0.294	0.149	0.051
	7				0.001	0.008	0.031	0.090	0.198	0.336	0.383	0.279
	8					0.001	0.004	0.017	0.058	0.168	0.430	0.663

TABLE II   BINOMIAL PROBABILITIES (cont.)

n	x	0.050	0.100	0.200	0.300	*p* 0.400	0.500	0.600	0.700	0.800	0.900	0.950
9	0	0.630	0.387	0.134	0.040	0.010	0.002					
	1	0.299	0.387	0.302	0.156	0.060	0.018	0.004				
	2	0.063	0.172	0.302	0.267	0.161	0.070	0.021	0.004			
	3	0.008	0.045	0.176	0.267	0.251	0.164	0.074	0.021	0.003		
	4	0.001	0.007	0.066	0.172	0.251	0.246	0.167	0.074	0.017	0.001	
	5		0.001	0.017	0.074	0.167	0.246	0.251	0.172	0.066	0.007	0.001
	6			0.003	0.021	0.074	0.164	0.251	0.267	0.176	0.045	0.008
	7				0.004	0.021	0.070	0.161	0.267	0.302	0.172	0.063
	8					0.004	0.018	0.060	0.156	0.302	0.387	0.299
	9						0.002	0.010	0.040	0.134	0.387	0.630
10	0	0.599	0.349	0.107	0.028	0.006	0.001					
	1	0.315	0.387	0.268	0.121	0.040	0.010	0.002				
	2	0.075	0.194	0.302	0.233	0.121	0.044	0.011	0.001			
	3	0.010	0.057	0.201	0.267	0.215	0.117	0.042	0.009	0.001		
	4	0.001	0.011	0.088	0.200	0.251	0.205	0.111	0.037	0.006		
	5		0.001	0.026	0.103	0.201	0.246	0.201	0.103	0.026	0.001	
	6			0.006	0.037	0.111	0.205	0.251	0.200	0.088	0.011	0.001
	7			0.001	0.009	0.042	0.117	0.215	0.267	0.201	0.057	0.010
	8				0.001	0.011	0.044	0.121	0.233	0.302	0.194	0.075
	9					0.002	0.010	0.040	0.121	0.268	0.387	0.315
	10						0.001	0.006	0.028	0.107	0.349	0.599
11	0	0.569	0.314	0.086	0.020	0.004						
	1	0.329	0.384	0.236	0.093	0.027	0.005	0.001				
	2	0.087	0.213	0.295	0.200	0.089	0.027	0.005	0.001			
	3	0.014	0.071	0.221	0.257	0.177	0.081	0.023	0.004			
	4	0.001	0.016	0.111	0.220	0.236	0.161	0.070	0.017	0.002		
	5		0.002	0.039	0.132	0.221	0.226	0.147	0.057	0.010		
	6			0.010	0.057	0.147	0.226	0.221	0.132	0.039	0.002	
	7			0.002	0.017	0.070	0.161	0.236	0.220	0.111	0.016	0.001
	8				0.004	0.023	0.081	0.177	0.257	0.221	0.071	0.014
	9				0.001	0.005	0.027	0.089	0.200	0.295	0.213	0.087
	10					0.001	0.005	0.027	0.093	0.236	0.384	0.329
	11							0.004	0.020	0.086	0.314	0.569
12	0	0.540	0.282	0.069	0.014	0.002						
	1	0.341	0.377	0.206	0.071	0.017	0.003					
	2	0.099	0.230	0.283	0.168	0.064	0.016	0.002				
	3	0.017	0.085	0.236	0.240	0.142	0.054	0.012	0.001			
	4	0.002	0.021	0.133	0.231	0.213	0.121	0.042	0.008	0.001		
	5		0.004	0.053	0.158	0.227	0.193	0.101	0.029	0.003		
	6			0.016	0.079	0.177	0.226	0.177	0.079	0.016		
	7			0.003	0.029	0.101	0.193	0.227	0.158	0.053	0.004	
	8			0.001	0.008	0.042	0.121	0.213	0.231	0.133	0.021	0.002
	9				0.001	0.012	0.054	0.142	0.240	0.236	0.085	0.017
	10					0.002	0.016	0.064	0.168	0.283	0.230	0.099
	11						0.003	0.017	0.071	0.206	0.377	0.341
	12							0.002	0.014	0.069	0.282	0.540

TABLE III   NORMAL CURVE AREAS BETWEEN 0 AND z

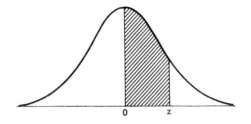

z	0.00	0.01	0.02	0.03	0.04	0.05	0.06	0.07	0.08	0.09
0.0	.0000	.0040	.0080	.0120	.0160	.0199	.0239	.0279	.0319	.0359
0.1	.0398	.0438	.0478	.0517	.0557	.0596	.0636	.0675	.0714	.0753
0.2	.0793	.0832	.0871	.0910	.0948	.0987	.1026	.1064	.1103	.1141
0.3	.1179	.1217	.1255	.1293	.1331	.1368	.1406	.1443	.1480	.1517
0.4	.1554	.1591	.1628	.1664	.1700	.1736	.1772	.1808	.1844	.1879
0.5	.1915	.1950	.1985	.2019	.2054	.2088	.2123	.2157	.2190	.2224
0.6	.2257	.2291	.2324	.2357	.2389	.2422	.2454	.2486	.2517	.2549
0.7	.2580	.2611	.2642	.2673	.2704	.2734	.2764	.2794	.2823	.2852
0.8	.2881	.2910	.2939	.2967	.2995	.3023	.3051	.3078	.3105	.3133
0.9	.3159	.3186	.3212	.3238	.3264	.3289	.3315	.3340	.3365	.3389
1.0	.3413	.3438	.3461	.3485	.3508	.3531	.3554	.3577	.3599	.3621
1.1	.3643	.3665	.3686	.3708	.3729	.3749	.3770	.3790	.3810	.3830
1.2	.3849	.3869	.3888	.3907	.3925	.3944	.3962	.3980	.3997	.4015
1.3	.4032	.4049	.4066	.4082	.4099	.4115	.4131	.4147	.4162	.4177
1.4	.4192	.4207	.4222	.4236	.4251	.4265	.4279	.4292	.4306	.4319
1.5	.4332	.4345	.4357	.4370	.4382	.4394	.4406	.4418	.4429	.4441
1.6	.4452	.4463	.4474	.4484	.4495	.4505	.4515	.4525	.4535	.4545
1.7	.4554	.4564	.4573	.4582	.4591	.4599	.4608	.4616	.4625	.4633
1.8	.4641	.4649	.4656	.4664	.4671	.4678	.4686	.4693	.4699	.4706
1.9	.4713	.4719	.4726	.4732	.4738	.4744	.4750	.4756	.4761	.4767
2.0	.4772	.4778	.4783	.4788	.4793	.4798	.4803	.4808	.4812	.4817
2.1	.4821	.4826	.4830	.4834	.4838	.4842	.4846	.4850	.4854	.4857
2.2	.4861	.4864	.4868	.4871	.4875	.4878	.4881	.4884	.4887	.4890
2.3	.4893	.4896	.4898	.4901	.4904	.4906	.4909	.4911	.4913	.4916
2.4	.4918	.4920	.4922	.4925	.4927	.4929	.4931	.4932	.4934	.4936
2.5	.4938	.4940	.4941	.4943	.4945	.4946	.4948	.4949	.4951	.4952
2.6	.4953	.4955	.4956	.4957	.4959	.4960	.4961	.4962	.4963	.4964
2.7	.4965	.4966	.4967	.4968	.4969	.4970	.4971	.4972	.4973	.4974
2.8	.4974	.4975	.4976	.4977	.4977	.4978	.4979	.4979	.4980	.4981
2.9	.4981	.4982	.4982	.4983	.4984	.4984	.4985	.4985	.4986	.4986
3.0	.4987	.4987	.4987	.4988	.4988	.4989	.4989	.4989	.4990	.4990

TABLE IV   NORMAL CURVE AREAS BEYOND z

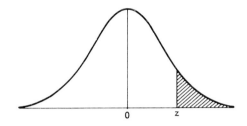

z	0.00	0.01	0.02	0.03	0.04	0.05	0.06	0.07	0.08	0.09
0.0	.5000	.4960	.4920	.4880	.4840	.4801	.4761	.4721	.4681	.4641
0.1	.4602	.4562	.4522	.4483	.4443	.4404	.4364	.4325	.4286	.4247
0.2	.4207	.4168	.4129	.4090	.4052	.4013	.3974	.3936	.3897	.3859
0.3	.3821	.3783	.3745	.3707	.3669	.3632	.3594	.3557	.3520	.3483
0.4	.3446	.3409	.3372	.3336	.3300	.3264	.3228	.3192	.3156	.3121
0.5	.3085	.3050	.3015	.2981	.2946	.2912	.2877	.2843	.2810	.2776
0.6	.2743	.2709	.2676	.2643	.2611	.2578	.2546	.2514	.2483	.2451
0.7	.2420	.2389	.2358	.2327	.2296	.2266	.2236	.2206	.2177	.2148
0.8	.2119	.2090	.2061	.2033	.2005	.1977	.1949	.1922	.1894	.1867
0.9	.1841	.1814	.1788	.1762	.1736	.1711	.1685	.1660	.1635	.1611
1.0	.1587	.1562	.1539	.1515	.1492	.1469	.1446	.1423	.1401	.1379
1.1	.1357	.1335	.1314	.1292	.1271	.1251	.1230	.1210	.1190	.1170
1.2	.1151	.1131	.1112	.1093	.1075	.1056	.1038	.1020	.1003	.0985
1.3	.0968	.0951	.0934	.0918	.0901	.0885	.0869	.0853	.0838	.0823
1.4	.0808	.0793	.0778	.0764	.0749	.0735	.0721	.0708	.0694	.0681
1.5	.0668	.0655	.0643	.0630	.0618	.0606	.0594	.0582	.0571	.0559
1.6	.0548	.0537	.0526	.0516	.0505	.0495	.0485	.0475	.0465	.0455
1.7	.0446	.0436	.0427	.0418	.0409	.0401	.0392	.0384	.0375	.0367
1.8	.0359	.0351	.0344	.0336	.0329	.0322	.0314	.0307	.0301	.0294
1.9	.0287	.0281	.0274	.0268	.0262	.0256	.0250	.0244	.0239	.0233
2.0	.0228	.0222	.0217	.0212	.0207	.0202	.0197	.0192	.0188	.0183
2.1	.0179	.0174	.0170	.0166	.0162	.0158	.0154	.0150	.0146	.0143
2.2	.0139	.0136	.0132	.0129	.0125	.0122	.0119	.0116	.0113	.0110
2.3	.0107	.0104	.0102	.0099	.0096	.0094	.0091	.0089	.0087	.0084
2.4	.0082	.0080	.0078	.0075	.0073	.0071	.0069	.0068	.0066	.0064
2.5	.0062	.0060	.0059	.0057	.0055	.0054	.0052	.0051	.0049	.0048
2.6	.0047	.0045	.0044	.0043	.0041	.0040	.0039	.0038	.0037	.0036
2.7	.0035	.0034	.0033	.0032	.0031	.0030	.0029	.0028	.0027	.0026
2.8	.0026	.0025	.0024	.0023	.0023	.0022	.0021	.0021	.0020	.0019
2.9	.0019	.0018	.0018	.0017	.0016	.0016	.0015	.0015	.0014	.0014
3.0	.0013	.0013	.0013	.0012	.0012	.0011	.0011	.0011	.0010	.0010

TABLE V   RANDOM NUMBERS

02946	96520	81881	56247	17623	47441	27821	91845
85697	62000	87957	07258	45054	58410	92081	97624
26734	68426	52067	23123	73700	58730	06111	64486
47829	32353	95941	72169	58374	03905	06865	95353
76603	99339	40571	41186	04981	17531	97372	39558
47526	26522	11045	83565	45639	02485	43905	01823
70100	85732	19741	92951	98832	38188	24080	24519
86819	50200	50889	06493	66638	03619	90906	95370
41614	30074	23403	03656	77580	87772	86877	57085
17930	26194	53836	53692	67125	98175	00912	11246
24649	31845	25736	75231	83808	98997	71829	99430
79899	34061	54308	59358	56462	58166	97302	86828
76801	49594	81002	30397	52728	15101	72070	33706
62567	08480	61873	63162	44873	35302	04511	38088
49723	15275	09399	11211	67352	41526	23497	75440
42658	70183	89417	57676	35370	14915	16569	54945
65080	35569	79392	14937	06081	74957	87787	68849
02906	38119	72407	71427	58478	99297	43519	62410
75153	86376	63852	60557	21211	77299	74967	99038
14192	49525	78844	13664	98964	64425	33536	15079
32059	11548	86264	74406	81496	23996	56872	71401
81716	80301	96704	57214	71361	41989	92589	69788
43315	50483	02950	09611	36341	20326	37489	34626
27510	10769	09921	46721	34183	22856	18724	60422
81782	04769	36716	82519	98272	13969	12429	03093
19975	48346	91029	78902	75689	70722	88553	83300
98356	76855	18769	52843	64204	95212	31320	03783
29708	17814	31556	68610	16574	42305	56300	84227
88014	27583	78167	25057	93552	74363	30951	41367
94491	19238	17396	10592	48907	79840	34607	62668
56957	05072	53948	07850	42569	82391	20435	79306
50915	31924	80621	17495	81618	15125	48087	01250
49631	93771	80200	84622	31413	33756	15218	81976
99683	58162	45516	39761	77600	15175	67415	88801
86017	20264	94618	85979	42009	78616	45210	73186
77339	64605	82583	85011	02955	84348	46436	77911
61714	57933	37342	26000	93611	93346	71212	24405
15232	48027	15832	62924	11509	95853	02747	61889
41447	34275	10779	83515	63899	30932	90572	98971
23244	43524	16382	36340	73581	76780	03842	64009
53460	83542	25224	70378	49604	14809	12317	78062
53442	16897	61578	05032	81825	76822	87170	77235
55548	19096	04130	23104	60534	44842	16954	99466
18185	69329	02340	63111	41788	74409	76177	55519
02372	45690	38595	23121	73818	74454	02371	94693
51715	35492	61371	87132	81585	55439	98095	55578
24717	16786	42786	86985	21858	39489	39251	70450
78022	32604	87259	93702	99438	68184	62119	20229
35995	08275	62405	43313	03249	74135	43003	63132
29192	86922	31908	42703	59638	31226	89860	45191

BIBLIOGRAPHY

Adams, J. K. *Basic Statistical Concepts*. New York, McGraw-Hill Book Co., 1955.

Burington, R. S., and May, D. C. Jr. *Handbook of Probability and Statistics with Tables*. Sandusky, Ohio, Handbook Publishers, Inc., 1953.

Clark, C. E. *An Introduction to Statistics*. New York, John Wiley & Sons, Inc., 1953.

Coxton, Frederick E., Cowden, Dudley J., and Bolch, Ben W. *Practical Business Statistics*. 4th ed. Englewood Cliffs, N.J., Prentice-Hall, 1969.

Cramér, H. *Mathematical Methods of Statistics*. Princeton, N.J., Princeton University Press, 1946.

Dixon, W. J., and Massey, F. J. Jr. *An Introduction to Statistical Analysis*. New York, McGraw-Hill Book Co., 1951.

Dornbusch, S. M., and Schmid, C. F. *A Primer of Social Statistics*. New York, McGraw-Hill Book Co., 1955.

Downie, N. M., and Heath, R. W. *Basic statistical methods*. 3rd ed. New York, Harper and Row, 1970.

Feller, William. *An Introduction to Probability Theory and Its Applications*. 3d ed. Volume I. New York, John Wiley & Sons, 1968.

Freeman, L. C. *Elementary applied statistics: For students in behavioral science*. New York, John Wiley and Sons, Inc., 1965.

Freund, J. E. *Modern Elementary Statistics*. New York, Prentice-Hall, Inc., 1952.

Fryer, H. C. *Elements of Statistics*. New York, John Wiley & Sons, Inc., 1954.

Games, P., and Klare, G. *Elementary statistics: Data analysis for the behavioral sciences*. New York, McGraw-Hill, 1967.

Glass, G., and Stanley, J. *Statistical methods in education and psychology*. Englewood Cliffs, N.J., Prentice-Hall, 1970.

Goedicke, V. *Introduction to the Theory of Statistics*. New York, Harper & Brothers, 1953.

Guilford, J. P. *Fundamental statistics in psychology and education*. 4th ed. New York, McGraw-Hill, 1965.

Hoel, Paul G. *Elementary Statistics*. 3d ed. New York, Wiley, 1971.

Hoel, Paul G. *Introduction to Mathematical Statistics*. New York, John Wiley & Sons, Inc., 1954.

Hogg, Robert V., and Craig, Allen T. *Introduction to Mathematical Statistics*. 3d ed. New York, Macmillan, 1970.

Huff, Darrell. *How to Lie with Statistics*. New York, Norton, 1965.

Klugh, H. *Statistics: The essentials for research*. New York, Wiley, 1970.

Kraft, Charles H., and van Eeden, Constance. *A Nonparametric Introduction to Statistics*. New York, Macmillan, 1968.

Mason, R. D. *Statistical techniques in business and economics.* Revised ed. Homewood, Ill., Richard D. Irwin, 1970.

Mode, E. B. *Elements of Statistics.* New York, Prentice-Hall, Inc., 1951.

Mood, A. M. *Introduction to the Theory of Statistics.* New York, McGraw-Hill Book Co., 1950.

Moroney, M. J. *Facts from Figures.* Baltimore, Md., Penguin Books, Inc., 1956.

Mosteller, Frederick, and Rourke, Robert E. *Sturdy Statistics.* Reading, Mass., Addison-Wesley, 1973.

Mosteller, Frederick, Rourke, Robert E. K., and Thomas, George B., Jr. *Probability with Statistical Applications.* 2d ed. Reading, Mass., Addison-Wesley, 1970.

Neter, John, and Wasserman, William. *Fundamental Statistics for Business and Economics.* 2d ed. Boston, Allyn and Bacon, 1961. Chapters 4 and 5.

Neyman, J. *First Course in Probability and Statistics.* New York, Henry Holt & Co., Inc., 1950.

Parzen, Emanuel. *Modern Probability Theory and Its Applications.* New York, John Wiley & Sons, 1960.

Savage, I. Richard. *Statistics: Uncertainty and Behavior.* Boston, Houghton Mifflin, 1968.

Shao, S. P. *Statistics for business and economics.* 2nd ed. Columbus, Ohio, Chas. E. Merrill, 1972.

Snedecor, George W., and Cochran, William G. *Statistical Methods.* 6th ed. Ames, Iowa, Iowa State University Press, 1967.

Spiegel, Murray R. *Theory and Problems of Statistics,* New York, Schaum Publishing Co., 1961.

Tukey, John W. *Exploratory Data Analysis.* Reading, Mass., Addison-Wesley, 1977.

Walker, H. M., and Lev., J. *Statistical Inference.* New York, Henry Holt & Co., Inc., 1953.

Whitney, D. R. *Elements of Mathematical Statistics.* New York, Henry Holt & Co., Inc., 1958.

Wilks, S. S. *Elementary Statistical Analysis.* Princeton, N.J., Princeton University Press, 1948.

BOOKS BY ENDERS A. ROBINSON

1954   Predictive Decomposition of Time Series with Applications to Seismic Exploration

1959   An Introduction to Infinitely Many Variates

1962   Random Wavelets and Cybernetic Systems

1967   Statistical Communication and Detection with special reference to Digital Data Processing of Radar and Seismic Signals

1967   Multichannel Time Series Analysis with Digital Computer Programs

1967   Forecasting on a Scientific Basis (with H. Wold, G. Orcutt, D. Suits, P. de Wolf)

1969   The Robinson-Treitel Reader (3 editions) (with S. Treitel)

1978   Digital Signal Processing and Time Series Analysis (with M. T. Silvia)

1979   Deconvolution of Geophysical Time Series in the Exploration for Oil and Natural Gas (with M. T. Silvia)

1979   Digital Foundations of Time Series Analysis
       Volume 1. The Box-Jenkins Approach (with M. T. Silvia)

1980   Geophysical Signal Analysis (with S. Treitel)

1980   Physical Applications of Stationary Time Series

1980   University Course in Digital Seismic Methods used in Petroleum Exploration

1981   Digital Foundations of Time Series Analysis
       Volume 2. Wave-Equation Space-Time Processing (with M. T. Silvia)

1981   Time Series Analysis and Applications

1981   Least Squares Regression Analysis in Terms of Linear Algebra

1981   Statistical Reasoning and Decision Making

INDEX

abscissa 2
acquisition phase 153
acts in payoff table 117
addition rule 25
addition rule for probabilities for
  incompatible events 20, 26, 28
advantageous game 22
a priori 55
arithmetic average 7
arithmetic mean 7
assessment of probabilities 121
attitude toward risk 130
average 7

Baruch 145
Bayesian decision making 101
Bayes rule 102
Bayes rule for flipping trees 101
Bayes theorem 102
betting on events 23
bettor's odds 23
bibliography 183
bimodal 7
bimodal distribution 42
binomial probabilities for
  $p = 0.5$ 177

central limit theorem 55, 59, 60
chance 16
classic definition of probability 15
collectively exhaustive 118
combination 41
common event 19
conditional analysis 115
conditional opportunity loss 129
conditional profit with perfect
  information 127
conditional value of perfect
  information (CVPI) 128
Conrad 159
contingent gain 21
contrary event 18

decision making 95
decision point 108, 109, 111
decision theory 98
decision under uncertainty 115
different parity 98
disadvantageous game 22
disjoint events 19
distribution phase 153
dual variables 87

E 69
empirical mean 44
empirical variance 45
empty event 18
entropy 79
entropy as a measure of
  uncertainty 83
equal and independent chance 52
equally likely 16
equally like case 16
equitable game 22
event 18
event in payoff table 117
expectation 20, 69
expectation of an event 21
expectation of several events 21
expected cost with perfect
  information 127
expected loss 107, 113
expected profit under certainty 128
expected profit with perfect
  information 127
expected monetary value 119, 120
expected opportunity loss
  (EOL) 130
expected utility 130, 134, 136
expected utility criterion 131
expected value 69
expected value of perfect information
  (EVPI) 127
expected value of sample mean 73

expected value of sample
  variance   75
experiment   15
experimental evidence   104

flipped tree   102
French roulette wheel   24
frequency   2
frequency distribution   2
frequency interpretation of
  probability   17
frequency polygon   4, 5

game   147
geometric mean   7
given that   47
Graunt   1
Greek alphabet   175
Greek letter mu μ 44
Greek letter sigma:
  lower case σ 45
  capital Σ 7

H   81
Halley   1
histogram   2, 56
hypothesis   95, 97
hypothesis testing   106

incompatible events   19
independence between individual
  births   46
interaction effect   91
interaction pattern   90
interference effect   91
interference pattern   90
intersection   19
investment   149
issues and charisma   88

Jaggers   160
judgmental weights   125

language   137

law of large numbers   16
level of significance   95, 97
Lichtenberg   145
life insurance   23
loss table   129

manipulation   152
mathematical expectation   20, 69
maximum entropy   79
mean   7
mean of binomial distribution   43, 44
measure of entropy H   81
median   7
mode   7
model building   147
model for the stock market   149
multiplication rule   29, 32
multiplication rule for
  probabilities   29, 30
mutually exclusive   118
mutually exclusive events   19

naive decision point   108, 109
normal curve   55, 59
normal distribution   55, 59
null event   18
null hypothesis   95
numerical data   1

ordinate   2

P   16
parameter   59
Pascal's triangle   39
payoff table   115
permutation   40, 41
personal definition of probability   18
point of inflection   60
population   51
population mean   69
population standard deviation   72
population variance   70
position and momentum   88
possible criteria   119

potential gain for bet   23
probability based on relative
  frequency   123
probability concepts in everyday
  language   139
probability α of a Type I error   111
probability β of a Type II error   111

random sample   52, 72
random sampling   52
random process   13
random variable   56
regret   107
regret at Type I error   107, 110
regret of Type II error   107, 110
relative cost   107
relative frequency   2, 56, 123
relative frequency definition of
  probability   16, 123
relative frequency distribution   2
review exercises   157
review problems   161

same parity   98
sample   52
sample mean   73
sample space   13
sample variance   8, 74
self-information   84
significance   95
significant   97
Sinclair   1
small investor   154
speculation   149
standard deviation   7, 11
standard deviation of binomial
  distribution   46
standard normal distribution   61

standard normal variable z 61
standard score z 11
stated position and inner feelings   89
statistical independence   14
statistics   1, 51
stock market speculation   147
subjective definition of
  probability   18

table of binomial coefficients   176
table of binomial probabilities   178,
  179
table of normal curve areas   180, 181
table of random numbers   182
theoretical frequency curve   53, 54,
  56
tree diagram   33, 101
trial of an experiment   42
true mean   69
Type I error   97, 107
Type II error   97, 107

unbiased estimate   75
uncertainty principle   87, 88, 90, 92
union   19
universal event   18
universal set   13
universe   13, 51
use of language   145
utility   131, 136
unimodal   7

value of information   127
variance   7, 8, 70
variance of binomial distribution   45
variance of sample mean   73

Wold   v.